The Conrey Placer Mining Company

The dredgemaster of Conrey No. 3 in his pilot house

THE CONREY PLACER MINING COMPANY

A Pioneer Gold-Dredging Enterprise in Montana, 1897-1922

Clark C. Spence

Montana Historical Society Press, Helena, 1989

Grateful acknowledgement is made to the Gilman family in Ruby, Montana, for the use of photographs for *The Conrey Placer Mining Company.* All photographs are from the Gilman family collection.

The Montana Historical Society Press, Helena, 59620

Printed in the United States of America
Composed in Baskerville by Arrow Graphics and Typography in Missoula, Montana. Printed by Thomson-Shore, Inc., in Dexter, Michigan.

Library of Congress Cataloging-in-Publication Data
Spence, Clark C.
The Conrey Placer Mining Company : a pioneer gold-dredging enterprise in Montana, 1897-1922 / by Clark C. Spence.
p. cm.
Includes bibliographical references.
ISBN 0-917298-18-7
1. Conrey Placer Mining Company—History. 2. Gold industry—Montana—Alder—History. 3. Gold dredging—Montana—Alder—History. I. Title.
HD9536.U54C667 1989
338.7'6223422'09786663—dc20 89-39452
CIP

To Vivian A. Paladin

Contents

	Preface	ix
Chapter 1	The Conrey and the World of Gold Dredging	3
Chapter 2	Of Men and Money	17
Chapter 3	An Era of Experimentation	29
Chapter 4	An Interlude of Steam	39
Chapter 5	The Triumph of Electricity	53
Chapter 6	Navigating Alder Gulch	69
Chapter 7	Of Men and Machines	87
Chapter 8	A Mine Is a Wasting Asset	113
	Notes	127
	Bibliographical Essay	155
	Index	157

Preface

This case study of the Conrey Placer Mining Company is at once corporate history, technological history, and local history. At the same time, it is working undredged ground. This was a pioneer gold-dredging operation, capitalized in a semi-private way by a relatively small number of investors without a public issue of stock. It was an innovative firm that was involved in the diffusion of a technology that gained national and international significance. If Montana was physically off the beaten path of American dredge development, Conrey was very much a part of the learning by doing—"the collective invention"— that accompanied the wedding of harbor and construction dredges and New Zealand mining dredges to produce a hybrid machine that came to be accepted as standard in American placer mining, as well as in the rest of the world. Even so, the path of technological adoption and adaptation was not always smooth: it proceeded by fits and starts, often by expensive trial and error, with investors paying for mistakes and experience.

In a sense, Conrey was a unique operation. It was the only operating dredge company in the country in which Harvard University—or any other university, for that matter—owned a controlling interest. By the terms of the bequest by which Gordon McKay's shares came to Harvard, the school enjoyed the income but could not dispose of them. First through Nathaniel Shaler, then through Henning Jennings, both with strong ties to the institution, Conrey for several decades paid its gold into Harvard coffers. Fortunately, Harvard kept the full set of records for posterity.

Sources are scattered, and I owe a number of debts for help received in locating materials and putting together this study. Robert Lovell of the Baker Library, Harvard Graduate School of Business, gave me free run of the Conrey Manuscripts in Boston. At the University of Wyoming, Gene Gressley settled me under the watchful eye of half a dozen Alfred Jacob Miller originals and made it easy for me to consult the papers of Hennen Jennings, one of the many remarkable collections on mining located at Laramie. I am also grateful to the Montana Historical Society in Helena for the use of newspapers, manuscripts, and books. At the offices of the Marion Power Shovel Division,

Dresser Industries, Marion, Ohio, Peter J. Gilewicz interrupted a busy schedule to be helpful beyond the call of duty. I am indebted to Lowell Gilman of Ruby, Montana, whose father, Leslie Gilman was a Conrey Placer Mining Company employee for more than a quarter of a century. Lowell Gilman now owns the old retort house and headquarters property and generously showed me the details of buildings, dredge maps, and the photograph album that is a treasure-trove of history so far as the company and its dredges are concerned. As graduate students, several people helped find elusive pieces of the puzzle over the years—especially Robert L. Spude and Ronald C. Brown. Errors are my own. I would also wish to thank the Department of History and the Research Board of the University of Illinois for financial support. All the while they thought I was skiing in Montana.

If I generally speak of the Conrey Placer Mining Company alone, rather than of it and its kindred concern, the Poor Farm Placer Mining Company, it is for the sake of convenience and simplicity. To be sure, between 1906 and 1915 two companies did officially exist; but their investors, their officers and their management were for all practical purposes identical, though technically electric dredges No. 3 and No. 4 were Poor Farm equipment during that period. Occasionally, I refer to the Poor Farm company when there is a legal agreement involved, but I have avoided any attempt to weave back and forth between the two firms. In any event, almost all of the business was done in the name of the Conrey Company.

The Conrey Placer Mining Company

Chapter One

The Conrey and the World of Gold Dredging

Among the many innovations in mining technology during the late nineteenth and early twentieth centuries, several would truly revolutionize the industry. Cyanidation and flotation would bring quiet but profound changes in milling and gold extraction. The low-grade copper techniques pioneered by Daniel C. Jackling and others would bring bulk handling and economies of scale to the working of previously unprofitable ores, just at a time when the new electrical and automotive industries were stepping up demands for the red metal. And gold dredging would apply the mass production of Henry Ford's America to placer deposits, again enabling the profitable working of ground that had never been worth turning over. It was the American West—California in particular—that would build on precedents set in New Zealand and Montana to develop a highly sophisticated dredge technology, one that would be disseminated in ever-widening concentric circles for use throughout the globe.[1]

Dredges are land-locked, self-contained, water-borne anomalies that dig their own pond as they move across the ground being mined. All recovery processes, except final retorting or refining, are conducted on the machine itself. Traditionally, dredges have been particularly effective in salvaging gold deposited by prehistoric streams or by glacial action. According to contemporary engineer Arthur Lakes, Jr., in 1909:

> A gold dredge consists of a floating hull with a superstructure, a digging ladder, endless chain of digging buckets, screening

> apparatus, gold-saving devices, pumps and stacker. It could be described as a floating mill with the addition of apparatus for excavating and elevating the ore.[2]

There were other types of dredges, of course—suction, dragline, and dry land among them—but the most conventional was the bucket ladder type described by Lakes. It became the most common, the most complex, and the most expensive of the dredge genre.[3]

Full-blown, bucket dredges were awesome and spectacular: ugly, graceless, megalosaurian. They might cost half a million dollars and be as large as an ocean freighter. Huge and tireless, they clanked relentlessly along, creating their own dirty pools and ripping pay dirt from bedrock fifty or even a hundred feet below the surface. These huge machines left an unsightly wake—"the kind of furrow that an enormous obscene un-house-broken worm might leave," in the words of John Gunther; or, as a more charitable observer expressed it, dredges "chewed the gravel like a boy extracting juice from an apple and then spat out the pulp, turning up the soil over the gold bearing area."[4] Unsightly and grotesque in an often scenic setting, bucket dredges stank of oil and muck and filth and scum. Around the clock, their greed seemed insatiable, their presence at once fascinating and dreadful.

If there was an air of romance about the ungainly gold dredge, it was because people were reminded of ocean-going vessels or steamboats on the Mississippi. Journalists and promoters wrote of the "Modern Gold Ship" and "Flagships of the Gold Fleet," or they likened the dredges to grain harvesters "Farming for Gold." Hard-nosed mining engineers warned against investing in a dredge labeled "gold ship," which was, they cautioned, "a term invented by a sophomoric reporter and employed by the incubators of wild-cats."[5] The dredge was "the acme of scientific mining and inventive genius," an instrument for conquering nature. It was "a wonderful modern invention," able to do the work of two thousand men and to make profits "where a cooly would starve." Because of this new marvel, a prominent monetary specialist suggested in 1903, the world would "not only be saturated with gold, but it would be nauseated with it."[6]

Basically, gold dredges gave miners a new way of looking at placer deposits. Applying mass production to a set of old problems, miners could work a maximum amount of gravel with steam, diesel, or electrically driven machinery, separating the gold aboard the hull and dumping the debris behind as the "boat" moved slowly forward. Because of the economies of scale—obtaining the greatest possible output in the highest possible working time and striving for the maximum possible extraction of gold—ground previously considered unprofitable or already worked over by less efficient techniques now became worth the trouble.

Enterprising Yankees had been tinkering with the idea of using dredges in California since 1850, when some of "the daring swashbucklers of the capitalist tribes" of Boston had outfitted the river steamer *Phoenix* as a gold dredger, only to find that the "Cumbrous arrangement . . . dredged more Money from the pockets of the owners than it did gold from the bed of the Yuba."[7] Subsequently, Californians took the lead in experimenting with an ingenious array of dredging apparatuses, ranging from diving bells and Dennison's "Ingurgitator Dredge" to a vacuum-operated "tube boat" of gargantuan proportions. All were failures and only fed the doubts of tough-minded American skeptics who dismissed gold dredging in the 1880s as "visionary" and "simply idiotic."[8]

Thus it was not California but New Zealand that was the source of the world's first successful gold dredges. It was not the Sacramento or the Yuba, but the swift-flowing, temperamental Clutha, the Shotover, and the Kawarau rivers that became "the kindergarten in which the early lessons were learned, dredgemasters were trained, dredge-builders were taught, and all the tricks and intricacies of the industry were developed."[9]

During the early 1860s, ingenious New Zealanders devised the primitive spoon dredge, a simple barge that dragged a "spoon" along the river bottom. The "spoon," usually made of buckskin and attached to an iron ring at the end of a long pole, could be winched up and emptied into a sluice either on the same hull or on an auxiliary pontoon towed behind. Such crude devices—at least ten were at work in Otago in 1865—caught on quickly and engendered great enthusiasm. "That dredging for gold will presently assume gigantic proportions there is no reason to doubt," a local editor predicted in the spring of 1869.[10] But the spoon dredge was highly inefficient, worked best in quiet waters, and was in the next decade superseded by the more complex current wheel dredge, a natural step in the evolution of the technology.

The current wheel dredge was a bucket-and-ladder-type machine operated by paddle wheels driven by a swift current. The bucket-and-ladder digging mechanisms had been borrowed from time-honored harbor dredges and adapted to water power. At the end of the 1870s especially, a number of current wheelers were built and gave good returns, but they were effective only where the river ran strong—seven or eight miles an hour along stretches of the Clutha, or Molyneux, as it was then called—and they lacked the power to dredge deep ground.[11] It was inevitable that by the early 1880s steam-driven bucket-and-ladder dredges would begin to replace current wheelers on the gold-producing rivers of New Zealand.

The first of the new steam dredges, Scott's converted current wheeler and Charles McQueen's modern iron-hulled leviathan with a bucket line on each side, made their debuts in 1882 on the Molyneux. Their use touched off a flurry of activity, which meant prosperous times for such Dunedin dredge builders as Kincaid, McQueen & Company, Cutten & Company, and R. S. Sparrow & Company.[12] By March, 1889, forty-nine dredges were at work in New Zealand, almost all of them on the South Island; but some were active in the west as well as in Otago Province, where only two of the thirteen were current wheelers.[13] From 1889 to 1892, in the midst of the "long depression" that lasted from 1865 until 1895, South Island enjoyed its first dredge boom—"a shaft of light on the clouded economic scene."[14]

If the 1890s were years of experimentation prior to another surge at the turn of the century, New Zealander engineers led the way. The number of dredges rose steadily: 36 in 1893; 84 in 1895, the year the first successful American gold dredge was launched; and 236 by 1900, the vast majority of them steam-powered and 80 per cent of them in Otago.[15] After numerous efforts, the New Zealanders discarded efforts to use the suction dredge and focused on improving the bucket-line machine. Through practical experience, these engineers strengthened the bucket chain, devised an elevator stacker to pile the tailings behind, applied special steel alloys to the bucket lips, used cocoa matting in the sluice to capture the gold, added special clamps to handle large boulders, built heavy and massive lower tumblers, and worked out techniques for dredging as deep as fifty feet.[16]

Detailed accounts in leading technical journals, papers read at professional meetings, and the kind of transfer of information endemic to a calling as nomadic as mining engineering made advances in the Southern Hemisphere well known in American mineral circles. By the mid-1890s, not only New Zealand dredges but also skilled technicians were being exported to the far corners of the mining world. One of them, Robert H. Postlethwaite, manager of the New Zealand Engineering Company of Dunedin, would play a leading role in the early development of the American dredge-mining industry. Migrating to California, in 1896 Postlethwaite became the chief dredge designer for the Risdon Iron Works in San Francisco, a pioneering mining machinery firm in the United States and one of the most important during the early years of dredge building.[17]

Postlethwaite's first Risdon rig was steam-powered with three-foot open-connected buckets (that is, with spaces between them). The dredge went to work at Park's Bar on the Yuba River near Smartsville in 1897, but it soon sank in the stream's turbulent waters. A year later, Wendell P. Hammon, an orchardman at Biggs, California, and Thomas Couch of Butte, Montana, put a second New Zealand-type Risdon on property at Oroville to good advantage, although it was clear that the technology needed strengthening and refining in order to succeed in the California environment. Even so, other influences were at work. Both Hammon and Couch had been in touch with men familiar with dredge developments in Montana and with eastern firms engaged in gravel, canal, and jetty work, always with an eye to improving their primitive technology.[18]

It was in Montana that a combination of local citizens and Chicagoans familiar with digging equipment introduced the *Fielding L. Graves*, the first successful American connected-bucket dredge. Christened by the Gold Dredging Company "before a large concourse of people" at Bannack, Montana, in May 1895, the boat slipped into the waters of Grasshopper Creek to the cheers of the crowd, inaugurating "the Dawn of a New Era of Placer Mining Methods," as a Dillon editor prophetically put it.[19] No doubt the *Graves* owed something to the New Zealand dredges that had been publicized since the early 1880s, but to a considerable extent the design was based on the bucket-elevator dredges that had been used for a number of years in building the Chicago drainage canal. The idea of applying dredges to gold recovery, it was said, began through "casual conversation in a railroad car, between men who never saw a placer mine in operation."[20]

Samuel S. Harper, an adopted Montanan who was described as "a man of money and vision" and not unacquainted with western mining, had been intrigued with a Bucyrus Company machine he saw at work on the new canal in Chicago. During a visit to Bannack, Montana, he discussed the dredge and an idea for adapting it for use in the placer fields with George Metlin, "a rosy-cheeked youngster just out of an eastern mining school." Harper then proceeded to acquire options on seventy-five acres of placer ground on Grasshopper Creek owned by Fielding L. Graves and Augustus F. Graeter. Having completed limited testing of the property, Harper returned to the Windy City, where he interested several men in the project, including Herman J. Reiling, a member of a leading Chicago contracting firm, and Benjamin Revett, who was experienced in Colorado hydraulic mining. Harper and Reiling took the lead in organizing the Gold Dredging Company, with mainly Chicago investors, including several former U.S. congressmen. They commissioned the Bucyrus Company of South Milwaukee to construct the new path-breaking electric dredge, which Harper played a major role in designing.[21]

Although the electrically powered *Graves* proved too light and many problems prompted the firm to have it rebuilt within a year, it was the pioneer in Montana and in the West at large. The limited successes of the *Graves* were evident and highly publicized: "The Gold Dredge Company has a Klondike of its own at Bannack," remarked the State Deputy Mine Inspector in 1897.[22] Four other steam-driven dredges would be built on Grasshopper Creek before the end of the century—the *A. F. Graeter,* the *Maggie A. Gibson,* the *Bon Accord* and the Coast Company dredge—all of which would have to be modified and reconstructed in short order and could be considered part of the learning process of American dredge building.[23]

In time and with much experimentation, the New Zealand and the Montana Grasshopper Gulch approaches to dredge construction produced a hybrid California type that was vastly superior to either. The cumbersome double-lift arrangement of the Bannack boats gave way to the single-lift system of Postlethwaite's earlier design, in which the bucket line carried the gravel to its highest point, eliminating the need for pumping the fine material an additional distance for screening or washing. Close-connected buckets of special steel were substituted for the open-connected type that Risdon had championed, so that, linked one to another, they gathered material loosened and dropped by previous buckets. The four-sided upper tumblers, which served as a huge sprocket to power the bucket line, now became pentagonal, hexagonal, or even round, and eventually were cast in one piece, body and shaft together, weighing as much as twenty tons. Although smaller, the lower tumblers that kept the line in place at the bottom of the digging ladder also evolved with a new type of steel alloy flanges and bearings for sustained underwater use under conditions of high abrasion. First high-carbon, then chrome, vanadium, and especially manganese steel combinations were used to lengthen the life of parts subject to heavy wear. In fact, dredging was the first practical application of Sir Robert Hadfield's discovery of manganese steel in 1882. The digging mechanism was developed more rapidly than the gold-processing equipment; but time brought improved scrubbing and screening devices, huge revolving screens ("trommels") washed by jets of water. The double banks of gold-saving tables over which the fines were sluiced, which had replaced the New Zealand shaking screens, remained the most primitive part of the apparatus.[24]

Because of the need to dig tough terrain that was often littered with boulders, the California-type dredges had to be heavier and stronger. These dredges abandoned the older system of headlines to keep their position and relied instead on steel-pointed spuds, massive posts dropped into the soil that provided a pivot around which the dredge could swing "like a dancer turning on her toe." Using one spud as an axis stuck in the bottom of the pond and manipulating the pressure on shore lines, the dredges chewed up the ground in an arc, see-sawing along. As the dredge pivoted, its buckets removed a cut of gravel, its width depending on the size of the bucket line, and the ladder was lowered slightly after each swing around the spud.

To discharge the leavings—or tailings—the belt stacker became standard, except on the Seward Peninsula in Alaska, although the early Montana machines had used flumes and the Risdons used a bucket conveyor to handle the debris. At first, the wooden hulls of the dredges were beautifully crafted by trained ship-builders and had a life expectancy of ten years, barring dry rot, but as timber costs rose and as dredges were exported to the tropics and the frozen North, steel hulls were used.[25]

Because of its low cost and the versatility of variable speed motors, electricity was the preferred source of power for dredges, either from individual generating plants or from local utility companies. Isolation often forced the use of steam or internal combustion engines, with diesel, gasoline, or distillate fuels hauled in by the barrel to remote areas like the Nome region in western Alaska.[26] In the field, engineers worked out techniques that enabled Montana dredges to operate through most of the winter, despite sub-zero temperatures. In the frozen North, thawing gravels by steam, electricity, or hot water proved to be too expensive in all but the richest ground, but a chance discovery in 1917 that cold water could be used advantageously eventually led to modern dredge fields covered with "bristling forests of thaw pipes, planted like thick-set, grotesque branchless trees." Alaskan dredges never managed to run through an entire winter, but in 1934 one of the monster rigs of the U.S. Smelting, Refining & Mining Company's Fairbanks Exploration Department worked 275 days of one season.[27]

By the turn of the century, it was clear that placer mining was about to be revolutionized. By 1900, 27 gold dredges were operating in the United States (16 in California, 6 in Idaho, and 5 in Montana), producing more than $520,000. In 1914, on the eve of World War I, there were 120 dredges in operation—60 in California, 42 in Alaska, 5 in Colorado, 5 in Montana, 4 in Idaho, and another 4 in other parts of the West. Of the $12,512,783 worth of gold dredged that year in the United States, California yielded more than half, $7,783,394; Alaska accounted for $2,340,000, and Montana was third highest at $835,615. In 1911, dredges produced 38.84 per cent of all gold mined and 85.32 per cent of all placer gold in California. Nine years later, the California dredge fleet was responsible for 51 per cent of all gold recovered and 98 per cent of all the placer gold mined in the state. Between 1903 and 1941, dredges were responsible for about 21 per cent of all Alaskan gold produced since 1880; at the industry's peak in the 1950s and 1960s, they were turning out as much as 82 per cent of the state's total output.[28] Nor was gold the only end product involved. The yellow metal was dominant in the beginning and was always important, but eventually there would be more tin dredges than any other kind. The huge machines would also be used to recover diamonds, sapphires, platinum, tungsten (scheelite), and such rarer minerals as osmium, titanium, thorium, tantalum, and columbium.

Dredging was a global industry, although American manufacturers predominated. Dredge machinery was normally made to specification by a heavy manufacturer like Risdon, Bucyrus, the Marion Steam Shovel Company, Western Engineering & Construction, or Yuba Manufacturing Company at Oroville, with the owner contracting privately for construction of the hull and installation.[29] But wherever they were built or wherever used, the California-type (sometimes called Yuba-type) dredges came to be accepted as the standard of excellence and were a significant export item. Of the world's 225 operating dredges at the end of 1915, 166 were in the United States and only 11 were not American-made. Already engineers like Charles Janin were serving as brokers, procuring used dredge machinery to ship abroad, for dismantling and installing the machinery in a new hull might save as much as half the original cost. Already Bucyrus dredges were at work at Kuala Lumpur and in the Bauchi district of Nigeria; those of the New York Engineering Company were digging at Chicksan, Korea, and in Siberia's Ochotsk Province; machines of the Union Construction Company of San Francisco were to be found in the tin regions of Portugal; Marion dredges plied the Kama River in the Urals; and the Yuba-type dredges worked ground from Paracale in the Philippines to the Pato Mountains of Columbia.[30] "The California-type dredge, known all over the world," one observer wrote in 1938, "is so efficient that it is being used on every continent where large quantities of low-grade metals are found."[31]

The heyday of American gold dredging was between 1905 and 1920, with California and Alaska occupying center stage and Idaho, Colorado, and Montana acting in secondary roles.

The giant dredges of the Guggenheims and the U.S. Smelting, Refining & Mining Company were already operating on a large scale in the far North when the price of gold increased from $20.67 to $35 an ounce during the New Deal years, touching off a new rash of dredging in 1934. But by federal order during World War II, gold mining was deemed an unessential industry and all of the big machines closed down except a handful that recovered gold as a secondary metal. Once the war was over, dredging was revived and continued until the mid-1960s, when most paying dredge grounds were worked out. Rising gold prices, however, might encourage a few efforts even in the 1980s and 1990s.[32]

But the first two decades of the twentieth century were the vital formative years and certainly the most exciting and exhilarating phase of dredge mining development in America. As the pre-eminent Montana placer firm of that period, the Conrey Placer Mining Company was very much a part of the revolution. The Conrey operation spanned the first twenty years of the century, showing a kind of continuity common only to a handful of multiple-dredge concerns. The corporation was formed at about the same time as American dredging commenced, and to some extent Conrey grew up with dredging. Initially, the company learned by experimentation and trial-and-error methods in its own peculiar environment, apart from the mainstream of California development, which it eventually embraced.

The question of the Conrey Company's relationship with other parts of the mineral West is a natural one. In general, how much innovation were the Conrey people responsible for, and how much interplay did they have with advancing dredge technology in California, Alaska, or elsewhere? In an advertisement in a Virginia City, Montana, newspaper just a few weeks before the last Conrey dredge closed down for good in 1922, the Montana Power Company lauded Conrey for its many dredge improvements, which had had an impact on the industry not only on the Pacific Coast but also as far away as the Philippines and South America.[33] During its early years, the Conrey Placer Mining Company was one of a small number of firms willing to undertake risk and innovation. It took experimentation and time for its technological efficiency to be translated into economic profit, but the company was acknowledged as one of the pioneers in gold dredging. Like many other companies, Conrey learned by doing, a not uncommon nineteenth century approach, but it was in the vanguard at least momentarily.

But Alder Gulch soon lost its early advantage. As visiting engineers were quick to point out, Montana was isolated from the mainstream of improvements in dredging and in some ways it lagged behind. Conrey dredges were the most advanced in the world when they first went into operation in 1898 and they were often improved, sometimes of necessity. But the time for diffusion of technology was when new dredges were being constructed, not when limited modifications of the old were being made. Dredge technology in California was advancing rapidly at the turn of the century, with new, improved machines being commissioned every week. After the first of its electrics, Conrey drew heavily on the California experience, but the company always made adaptations to fit its own circumstances—improvements that might be incorporated elsewhere.

When John Power Hutchins visited Alder Gulch in 1907, the internationally known dredge specialist observed that the bedrock there was a volcanic ash resembling the Oroville bottom but was much more tenacious and sticky, making it more difficult to scour the buckets than in California.[34] Hutchins was impressed with a Conrey device to prevent a loss of material spilled from the buckets—an arrangement using a stream of water to wash the material and save it for the tables. He was also aware that the company had experimented with bucket castings, using both annealed and unannealed castings in the same bucket chain and finding no appreciable difference in wear and strength. Conrey had also developed trommels with manganese steel baffle plates and rectangular holes set level with the spiral to advance the material through them. Hutchins commended Conrey's development of cast manganese steel wearing plates and friction rollers (ordinary car wheels) for driving them.[35] The company's consulting engineer, Hennen Jennings, listed several other Conrey improvements, including watertight bearings for the lower tumbler and the first use of the water rheostat and direct gear-connected drive motors to the bucket chain used with a friction clutch. Jennings also noted that

Conrey was a leader in experimenting with large buckets and the overhead or "aerial-spud" system, with lines to the stern and bow banks converging on a spindle-bearing with pulleys. When the boat moved forward, the bow lines were brought in and the stern lines loosened and clamped down.[36]

Conrey's experience with buckets, tumblers, and ladders seemed to Hutchins to have been separate from but similar to those of other dredging districts. There had been a progression from the 750-pound buckets of the *Maggie Gibson*, Conrey's first dredge, to the monster 5,000-pounders on No. 4 electric, its last dredge. From experimentation with various steel alloys, Conrey settled on locomotive tire steel pins and manganese steel bushings and bucket lips as most satisfactory. The company kept complete spare bucket chains on hand and treated each chain as a unit for repair and renewal to maintain uniform bucket pitch, thus preventing excessive wear from slippage on bucket bottoms and tumbler faces. Conrey also had experimented with teeth of different shapes, angles, and types on the bucket-cutting edge but found no advantage in digging over conventional teeth. Flax packing and engine oil lubrication of lower tumbler boxes and bearings on rollers proved especially effective in keeping the ladder functioning. Former Conrey manager Adolph H. Wiseman believed that the company's major contributions were in bucket line improvements and in the area of cold weather operations.[37]

Even though separated by distance from California, Conrey officials did make some effort to keep up with advances there. Engineers from the Oroville or Folsom fields visited the Ruby site occasionally, and Conrey superintendents or their assistants observed procedures in California. Company engineer Hennen Jennings sometimes brought his men in Montana abreast of current new trends and equipment. In mid-1917, he sent blueprints of a new type of high-pressure nozzle used by the Natomas Company in California to help clear the buckets of sticky clay, asking engineers at Ruby to give an opinion on its potential usefulness on Conrey dredges. The consensus there was that it was a useful apparatus but that the company had earlier devised a similar attachment of its own, using low instead of high pressure, which worked very well.[38] Jennings was also interested in jig experiments done by the Natomas people, but he decided that the results did not justify trying them in Montana.[39]

But in its semi-wilderness location at Ruby in southwest Montana, between Nevada City and Alder, the Conrey Placer Mining Company simply did not have easy access to the kind of information about dredging that owners and manufacturers were able to provide in California, where dredges were much more common. At the beginning of 1916, for example, there were eighty-five gold dredgers in California and only seven in Montana.[40] When engineer George J. Carr visited Ruby in 1919 to examine the Conrey property, he thought that the operation was well-managed but that some of the details were "probably not quite up to those practiced in the California fields." Because they had at least 50 per cent more riffle surface area, California dredges saved much more of the fine gold than Conrey did in Alder Gulch. The basic difference, Carr thought, was a matter of physical isolation: California dredge operators met every two months or so to discuss the best and most up-to-date methods.[41]

In many ways, the story of the Conrey Placer Mining Company is the story of the early years of the American gold dredging industry. Yet, before Conrey (and other mining operations) adopted the standard California-type machine, the firm went through its own period of trial-and-error, sometimes with positive results, sometimes not. In the end, even though its innovative expressions were tempered by developments on the Pacific Coast, Conrey had a separate and unusual history of its own.

Chapter Two

Of Men and Money

In the written histories of Montana, the Conrey Placer Mining Company is rarely mentioned-and then only in passing.[1] The company did nothing spectacular. It was not involved in a "War of the Gold Kings." It controlled no newspapers and made no attempt to dominate the state's political life. It experienced no devastating, knockdown fights with organized labor. It employed only one small-town lawyer and was involved in almost no litigation. Conrey was a relatively closed corporation, with stock owned by comparatively few investors, with its policies amicably determined without shareholder friction and implemented by managers of reasonable competence. The company was willing to experiment with new equipment and in the long haul, despite early misfortunes, it generally made profits. After 1903, at least, it had the distinction of being the only active mining company in the United States in which Harvard University owned a controlling interest. In the mining world, the Conrey Placer Mining Company was an unusual combination, guided by unusual people.

For nearly a quarter of a century, Conrey's dredges worked the ground at the lower end of Montana's celebrated Alder Gulch, through which flows the creek bearing the name of the white-barked tree so common in northern mountain regions. Alder Creek rises on the northern slopes of Old Baldy Mountain, at the south end of the Tobacco Root range, and flows northwest in an irregular curve to the Ruby River at Laurin. The rich gold discovered in Alder Gulch in 1863 had created a placer boom of classic proportions. Wells, Fargo & Company estimated that approximately $75 million was taken from a stretch of about ten miles, much of it steep and narrow; others believed these figures were low by about one-third. After the cream had been skimmed by pan, cradle, Long Tom, and sluices, the gulch was worked over again and in some places a third time by ground sluicing or even tunneling before the dredges arrived.

During the 1860s, Virginia City was the crown jewel of Alder Gulch. A bustling, hustling center of wealth, excitement, and crime typical of the mining frontier in general, the town joined with Bannack to purge its outlaws. With a population of near ten thousand in 1865, Virginia City became Montana's territorial capital for more than a decade, even though its placers had played out and its numbers dwindled after 1870. Sprawling down the gulch below the town like beads on a string lay Nevada City, Centerville, Adobetown, and Junction City, each camp with its own fleeting moment of glory based on placer gold and each leaving with the usual environmental havoc in its wake.

While the Conrey dredges would re-work some of the already mined ground with their "mass production" approach, the company first focused on the lower end of the gulch. The waters in that section of Alder Creek had formed a low alluvial fan, which because of its relatively low value and the high water level had been left untouched by earlier miners. It was there, near the towns of Alder and later Ruby, that six hundred acres or so seemed workable, despite an uneven bedrock, fairly large boulders, and the steep grade of the creek. The region was deemed to be late Pleistocene in time, perhaps contemporary with deposits in Colorado, but deposited over a shorter time span than the California dredging fields. The narrow gulch rarely measured more than three hundred feet across. Its surface had a 1 per cent slope to the west, with bedrock inclining slightly more than that in the same direction, so that as dredging progressed westward depth would increase until it reached a maximum of about sixty feet some two-and-a-half miles from the mouth of Alder Gulch. The gravel resembled the Oroville bottoms: at bedrock, it was volcanic ash, but was more tenacious because of an irregular distribution of clay and was rarely cemented like the hardest of the Oroville district. The gold itself, like that taken out in earlier years, tended to be more coarse than that found in the California dredge fields, although large nuggets were rare.[2]

A key figure in the formative years of the Conrey Placer Mining Company was Philip N. Moore, a mining engineer of considerable experience who had close ties to Boston capital. In an interview in 1922, Moore concluded: "Probably on the whole the most interesting enterprise with which I have been connected, and which in fact I organized and assembled, was the Conrey Placer Mining Company of Montana."[3] The son of a pioneer civil engineer engaged in railroad and canal building in the Mississippi Valley, Moore had been graduated in the classical course of studies at Miami University of Ohio. From 1870 to 1872, he had studied at the Columbia School of Mines, which was on its way to becoming the most important mining school in nineteenth century America. After half a dozen years with state geological surveys in Michigan, Missouri, and Kentucky, working with such important figures as Nathaniel S. Shaler, T. B. Brooks, and Raphael Pumpelly, Moore was drawn to the booming silver camp of Leadville, Colorado, where he joined an exciting group of clever young technical men who went on to brilliant careers in mining and smelting—Albert A. Blow, Max Boehmer, David W. Brunton, Arthur D. Foote, Ferdinand Van Zandt, and James B. Grant, among them. In Leadville, he supervised the operation of the camp's first smelter, then built a second, the LaPlata Works.[4]

From Leadville, Moore went to Kentucky to open the Slate Creek Iron Property. As chief executive and part owner for the next quarter of a century, he watched the concern expand and prosper until its sale in 1908. After 1889, he also operated as a mining consultant, traveling as many as fifty thousand miles a year out of St. Louis, blanketing the West as well as Canada and Mexico. It was in his capacity as a consulting engineer that he stumbled upon the property that would become the basis for the Conrey Placer Mining Company. Eventually, during World War I, when he developed the important Admiralty Zinc Company in Oklahoma, he became president of the American Institute of Mining Engineers and played an important role in coordinating government action on strategic minerals.[5]

In 1896, Moore was employed by Harry G. Bixby of Nashua, New Hampshire, and Henry M. Bigelow, of Brookline, Massachusetts, both capitalists with an office based in Boston and not averse to at least cautious investments in western mines.[6] That year, Bigelow and Bixby sent Moore to examine a placer prospect in Idaho. At the request of the promoter who accompanied him, the engineer stopped off at Alder Gulch to see the German Bar property, on which the promoter believed he held an option. The German Bar ground included about a mile and a half of the gulch, and the operation contemplated reworking the old tailings using machinery, a possibility that intrigued Moore. Unfortunately, the promoter did not have an option; but Moore was interested enough to ask the current option holder to offer the property to Bixby and Bigelow if he decided not to exercise his bond.

During the same visit, Montana miner John C. Sloss showed Moore the Phillip Conrey ranch, located a mile or so below the mouth of Alder Gulch. Born in Ireland, Conrey crossed the plains to Virginia City in 1864 and had apparently acquired the land years earlier in lieu of the payment of a $600 debt. The Conrey property "was considered one of the best hay ranches in the state," and Sloss knew it could be acquired for $30,000, more than three times its value for agricultural purposes.[7]

While Moore continued on to Idaho, the option lapsed on the German Bar property, and it was promptly taken up by Bixby and Bigelow. Moore re-visited Alder Gulch in October and closed negotiations on the German Bar. His clients paid $100,000 for the property, one-third down and the rest in three equal payments.[8] At the same time, the group began prospecting the Conrey ranch on which would be developed 200 acres of payable dredging ground, once they were acquired during the following year. Veteran placer miner John Sloss superintended the preliminary exploration of the Conrey property using test shafts and a Keystone drill; but recognition of the possibilities and the acquisition and early development of the property were due to the patient, even-handed Moore, who handled the engineering, while Harry Bixby tended the business side.[9]

Also affiliated with the Conrey enterprise from an early date was Nathaniel S. Shaler, professor at Harvard and a nationally known geologist. Shaler was one of a number of remarkable academics at the end of the nineteenth century who successfully combined teaching and research with investment in mining ventures and the professional examination of mining ground for vendors or buyers.[10]

A graduate of Lawrence Scientific School, where he had studied with Louis Agassiz, Shaler taught at Harvard for forty years and became a legend among undergraduates there. Over the years, thousands enrolled in his introductory Geology 4, the most popular class on campus, in part because it was considered a "snap" course requiring little work and in part because of the charm, wit, and personal magnetism of the wiry, tawny-bearded professor who taught it.[11] From 1891 until his death in 1906, Shaler was dean of Lawrence Scientific School and did much to shape that division of the university.[12]

Early in his career, Philip Moore had worked under Shaler's guidance for four years, while Shaler was State Geologist of Kentucky. Shaler became Moore's business associate and consultant. Some contend that it was Shaler's insight that was responsible for using test drilling to determine that the ground in Alder Gulch was a commercially feasible proposition, but actually the geologist never saw the property prior to its acquisition. At the time, Shaler was associated with Bigelow and Bixby, and he had accepted Moore's judgment on the property. He visited the ranch for the first time in the hot, dry summer of 1897, after the property had already been bought and test drilled.[13]

Another vital member of the early Conrey enterprise—one who provided the wherewithal for expansion—was Shaler's close friend and neighbor, millionaire Gordon McKay of Cambridge and Newport. A native of Pittsfield, Massachusetts, McKay early showed an aptitude for mechanics and engineering. After several years helping build the Western Railroad and enlarging the Erie Canal, he built his own machine shop to do specialized work for paper and saw mills; but he made his fortune through buying and improving patents for manufacturing shoes. McKay rode the crest of demands for boots and shoes during the Civil War, and by 1876 the 1,500 machines he controlled were bringing him half a million dollars annually in royalties. During his lifetime, McKay was sole or joint patentee of more than forty inventions, which brought him wealth estimated at $40 million.

McKay sold out his various interests in 1895 to the Goodyear Company, and at the time of the inception of the Conrey Placer operation, he was living quietly, devoting his time to philanthropic work, including the McKay Institute at Kingston, Rhode Island, which he established to educate black boys.[14]

The tall, striking, affable McKay had been friends with Shaler since 1865; they were close neighbors in Cambridge and were drawn together by a mutual interest in mechanical devices and in the challenges and complications of mining. Shaler later remarked that he never knew any man so well for so long. McKay followed Shaler's professional work and sought his advice on various enterprises and was always much impressed with the geologist's ability to predict and to meet unforseen difficulties. Apparently, McKay also sought Shaler's council on how to use his money for the public good. Twice married and divorced, McKay argued that no man ought to pass along more than $100,000 to his son; in the best Gospel of Wealth tradition, society in general should benefit. It was Shaler who broke down the millionaire's "serious dislike" of Harvard University, and in the end the bulk of McKay's fortune—"clean money," the geologist was glad to say—including his controlling interest in the Conrey enterprise, would go to benefit that institution.[15]

Late in December 1897, the Conrey Placer Mining Company was incorporated in Montana to run for forty years. Nominally, the company's capital was $50,000, in 500 shares of $100 each. Its capital stock actually subscribed at the time of chartering was $600, one share held by each director—Harry Bixby, Henry M. Bigelow, Philip Moore, Nathaniel Shaler, and Gordon McKay—plus one share held by Emilie Bixby, director Bixby's wife.[16] A series of agreements, made perhaps as early as October 1896 and involving the same parties in what was loosely called the Bixby Mining Association, created two trusts—"Trust #1" and "Trust #2"—that allowed Bixby and Shaler to act on behalf of the others to acquire and work mining property in the West. If Bixby were incapacitated or died, Shaler was empowered to act alone.[17] Officially, Conrey company offices were in Virginia City in the rooms of attorney William Alexander Clark, a lawyer of much less notoriety than his namesake, copper magnate William Andrews Clark of Butte. Clark, who was described as a "rather tough vigorous westerner, but easy to get along with," would serve for years as the firm's legal advisor. Conrey's directors also met in Boston, as they did in mid-February 1898, when they elected Gordon McKay president, Harry Bixby secretary-treasurer, and Philip Moore manager.[18]

It is not clear how much actual capital came into the company at this time, because many of the early transactions were apparently conducted through Trust #1 and Trust #2 and were in the form of both assessment and loan transactions. Records indicate that prior to January 1, 1899, Trust #1 paid out $35,105.33 to John C. Sloss, $1,405.50 to C. D. Gavin (both of the German Bar operation), $9,641.66 to the Montana Flume Company, $4,827.82 in expenses at German Bar, $6,263.84 to the Conrey Placer Mining Company, $7,000 to the County Poor Farm, and $5,000 to owners of property under option. At the same time, during a three-month period, McKay put $3,400 into Trust #2, Moore $2,560, Shaler $2,960, Bigelow $1,960 and Bixby $1,400 – a total of $12,547.[19] Special assignments were made for different purposes: Bigelow, McKay, and Mrs. Bixby were assessed $2,500 each for exploration; Moore, McKay, Shaler, Bigelow, and Mrs. Bixby were levied $1,160 each for acquiring an option on the Collins Ranch; Bixby, Bigelow, Moore, McKay, and Shaller were each assessed $1,400 for the option on the Van Brocklin property, while Mrs. Bixby's contribution was set at $334.[20] How much of the early capital came in the form of loans is hard to say, but it was enough to be noticeable. The Directors' Records of March 24, 1898, report: "Voted to ask of Mr. Gordon McKay the loan of $10,000, under the agreement with him." Entries also show substantial borrowing from Shaler, Moore, and Bigelow.[21]

The directors increased the company's capital to $200,000 in the summer of 1898, an action officially recorded early the following year.[22] In September 1900, at a time when the Conrey Placer Mining Company had issued $176,900 in stock, $80,000 of which was held by McKay, the company again voted to increase capital, this time to $300,000, in order to acquire the remaining property of the German Bar Mining Company. The expansion would issue 1,000 new, fully paid-up shares of $100 each to stockholders of the German Bar firm, the assets of which were by this time essentially in the hands of the Bixby Mining Association. The transfer was made to the Conrey Placer Mining

Company for $100,000; simultaneously, the directors authorized the issuance of $300,000 in fifteen-year gold mortgage bonds at 6 per cent interest.[23]

By June 1892, the company listed an indebtedness of $550,365.92, but it was in the process of expanding and restructuring its obligations. A month earlier, the directors had voted to authorize $800,000 in preferred shares, with the intent of exchanging some of them for mortgage bonds, although the full amount was not issued immediately and shareholders did not make the action official until 1905, when the certificate was registered with the Montana Secretary of State. But in September 1892, when plans were completed to convert the bonds into First Preferred Stock, McKay held $259,000 worth, Shaler $14,000, Moore $22,000, and Bigelow $5,000. At the same time, $231,287.54 in notes and interest advanced through the two trusts and the Conrey Company were to be replaced by Second Preferred Stock. Of these, McKay held $88,199.01; Philip Moore, $29,158.97; Ada Moore, his wife, $215.22; Henry Bigelow, $34,111.60; Harry Bixby, $4,618.33; and Mrs. Bixby, $49,487.35. In the exchange, McKay held 822 shares of Second Preferred Stock in his own name, with another 70 labeled "coll H. M. B.," 494 as "coll E. A. B.," and 46 as "coll H. G. B.," presumably holding these three packets as collateral for Bigelow and the two Bixbys. And finally, another 202 shares of Second Preferred Stock were issued to offset debts and interest to Bigelow (2), Moore (7), Shaler (3), and McKay (188, plus 2 held as collateral for the Bixbys).[24] In all of these capital expansions and issuances of preferred stock, the same investors were involved. These were in-house operations, with no offerings to the public, forming the capital foundations upon which the company based its early years of experimentation, including the construction of two steam dredges.

From the beginning, through the two trusts and then through the Conrey Placer Mining Company in its own name, Bixby acquired promising placer ground in Alder Gulch. By a deed of November 29, 1897, Philip and Sarah V. Conrey conveyed the ranch property that would give the company its name; included were some 410 acres, plus the Conrey and Kurfess placer mines, with all water and other rights. The price was $27,500, although an additional sum had probably been put down to obtain the option. Two years later, Conrey, acting on behalf of the estate of a minor, conveyed to Bixby thirty-three shares in the Montana Flume and Mining Company. In November and December, Bixby concluded agreements whereby Oscar A. Sedman of Virginia City deeded over his interest and rights of the German Bar Fluming Company and the Montana Fluming and Mining Company in exchange for $15,615.50. Less than two weeks later, W. W. and Mrs. S. B. Gregory of Pasadena, California, sold their Alder Gulch land to Bixby for $12,500. In the same month, Bixby acquired two pieces of placer ground between the German Bar Fluming Company and the Montana Flume and Mining Company from Henry and Mary Elling for $31,500.[25]

In the summer of 1898, Bixby acquired other interests in the Montana Fluming and Mining Company from David R. and Mary A. Beck of Missoula at a cost of $3,693.93. He bought 480 acres in Madison County from Fabien and Victoria Legris of Kansas at a price of $5,800, and he acquired two small placer claims in Ruby Canyon at the mouth of Davy Gulch from William Hinch and George T. Harrison at a cost of $200.[26] Bixby acted as trustee, with the various properties transferred to the German Bar Mining Corporation in the one case, to handle the German Bar and kindred ground, and to the Conrey Placer Mining Company in the other, to deal with the Conrey ranch and property farther down Alder Gulch.[27] But it took time to obtain all the parts of the Montana Fluming and Mining Company. Augustus Sweet of Butte agreed to surrender the last of his interest in that concern for $1,500 in mid-1901; Joseph H. and Elizabeth Carter ceded their interests more than a year later for $2,000.[28]

Gradually, additional property—some substantial, some minor—was added to the Conrey company's holdings. From the estate of Elias R. Jones, Gordon McKay paid $2,500—as the highest bidder—for Alder Gulch property, including a house, the McGovern claim, and "Morton's Ground." The property known as the "Sedman Place" or "Sedman House" came into Conrey hands in 1901 for $2,250. The company obtained a sliver of land from the Rodgers in 1902 for fifteen dollars and bought forty acres from George L. and Clara M. Ramsey of Helena.[29]

Late in 1903, Eliza J. Van Brocklin agreed to accept $10,000 from the company for the deed to a substantial piece of ground and an interest in the Stanley Ditch. Precisely what this interest was is not clear. The Conrey account journal for December 1901 lists $100,000 under real estate for the Van Brocklin option. A 1910 deed shows Eliza J. Van Brocklin, "widow," transferring 24.7 acres to the company for a nominal one dollar on the books.[30] Could it be that Van Brocklin was the major owner of the German Bar property, which supposedly sold for $100,000? Or was Eliza conveying another piece of ground to the company?

In October 1901, by agreement with the Madison County commissioners, the company paid $3,950 for the land between the north line of the Poor Farm and the county road, complete with water rights. Conrey had already purchased the 400-acre Collins Ranch and the 140-acre County Poor Farm, the latter at a cost of $7,000. Both properties were acquired in the name of Harry Bixby as trustee under the provisions of the December 31, 1897, agreement until the affiliated Poor Farm Placer Mining Company was organized in 1904 to take over the property.[31]

In 1906, the Conrey Company purchased an additional piece of Alder Gulch placer grounds from a Chinese group for $1,000. In several instances, the company acquired and subsequently patented unpatented placer claims. One—the Emilie (no doubt named after Mrs. Bixby)—was taken up by Shaler, Henry and Jeannette Bigelow, and Philip and Eva Moore. The claim was conveyed to the company and subsequently came into dispute when the Northern Pacific Railroad claimed it as a lieu selection for land already taken.[32] The final resolution of the dispute is not known.

By contract and option to purchase on the last day of 1897, Bixby as trustee of the German Bar Mining Company obtained the rights to the Elling property. The deed was transferred to the Conrey Placer Mining Company, although Henry Elling's heirs later had to re-deed the property in order to rectify a clerical error.[33]

Thus, Harry Bixby acquired properties under both his trusteeship and the Bixby Mining Association and assigned them to the German Bar Mining Company, of which he also served as trustee until the Conrey Placer Mining Company and the Poor Farm Mining Company were effectively organized. An undated memo concerning the German Bar Mining Company, probably written at the turn of the century, indicates that the group held a 95 per cent interest in the Montana Flume Company, the Conrey Ranch, and the Elling Ranch and an 80 per cent interest in the German Bar, but "on a/c of fraudulent representation by G——— he has forfeited his claim under contract to the 10%." "G———" may have been C. D. Gavin, listed in 1897 as a "mechanical engineer for the company" and as holding an interest in the property.[34]

All of this ground in Alder Gulch was gradually amassed to form an integrated, cohesive holding. Some was virginal soil, so far as mining was concerned, but much had already been worked by more primitive, labor-intensive methods. The portion of the creek bed that was profitable for dredging was much less than the whole. Eventually in 1912, when the Conrey operations were at their prime, it was reported that the company owned "practically the whole of Alder Gulch from Nevada, a mile below Virginia City, to Alder, a distance of eight miles, the only exception being a fraction of a claim at Junction, owned by James Vanderbeck." Within a few months, even the Vanderbeck property was in the fold.[35]

Chapter Three

An Era of Experimentation

As with mining technology in general, early gold dredging was a matter of trial and error, of adaptation, experimentation, and frustration. The Conrey placer operations were no different. From the beginning, the work of the Conrey group was a steady progression of experimentation—from large-scale sluicing, to the Lidgerwood cableway system, to the steam-driven Hayward "orange peel" shovels, and finally to the primitive bucket-ladder dredge—before a satisfactory approach was struck. Even dredging was at first a hit-and-miss kind of proposition, with a lot of technical adjustment and modification required to produce a profitable venture. As Conrey engineer Philip Moore remarked: "we built machinery and tore it to pieces, re-building stronger, and repeating the operation until at last machinery was evolved to meet our requirements." A knowledgeable Montana state official made a similar point when he lauded the Conrey group: "It took a daring initiative, a large amount of capital, extreme patience and much ingenuity to devise and elaborate the necessary methods and await the outcome."[1]

The first efforts were made on the German Bar property, acquired from the old German Bar Fluming Company, which had been sluicing near Junction about halfway up the gulch toward Virginia City. Apparently, owners J. C. Sloss and C. D. Gavin both came with the property, and the Conrey group at first continued the sluicing operation under a lease and bond agreement. Soon it would purchase the ground outright.[2]

The next approach at the German Bar property was experimental and was done under the supervision of Sloss, now general manager on the scene as well as a shareholder. The German Bar Mining Company began working Alder Gulch in 1897, with a new plant on German Bar, just below Adobetown, two-and-a-half miles down from Virginia City. It is not clear who designed the novel equipment, although variations of it had been tried in Alaska, Oregon, and northern California. Philip Moore was no doubt a key figure, and perhaps Gordon McKay or Nathaniel Shaler also had a hand in it. In any event, whoever it was had likely witnessed the digging of the Chicago drainage canal and had seen workers using the cableway dirt carriers manufactured by the Lidgerwood Manufacturing Company of New York, maker of steam and electric hoists and of cableways for mining, quarrying, timbering, or dam or canal construction. Also at work on the Chicago drainage canal was a variety of steam-powered shovels, among them Marions and Bucyruses, from companies that would later be prominent in the production of gold dredging equipment.[3]

The Lidgerwood Radial Cableway, as it was called, was an automatic digging apparatus featuring two towers and a cable-bucket assembly. One tower was stationary, and the gravel was brought to it, dumped, sized, and eventually sluiced. The other tower, the power source, was on rails and moved in a semi-circle around the fixed tower. The plant used a cable with a 400-foot span to draw a large Knight excavating bucket, raise it, and empty its contents into an elevated hopper before the process of separating the gold began. Carrying one and a half cubic yards of material per load, the bucket was controlled by a system of cables running between the central pivot tower and the smaller moveable Atwin-frame structure about 150 yards apart. The main, or hopper, tower was built of eight-by-eight timbers extending up through a large wooden hopper to a height of sixty-seven feet, with a pivot on top for main cable support and sheaves for operating ropes. Despite cables and guy wires, this ball-bearing bonnet could turn on its apex, allowing the traveling head tower to move through an arc of 180 degrees. As originally conceived, the smaller tower—the traveling head tower—was the power source, though there is some reason to believe that in the later phases of its use, the power plant was moved to a more stationary engine room near the hopper tower. The smaller tower was equipped with a powerful steam engine, a fifty-four-inch locomotive-style boiler, and cable anchorage; it moved on curved tracks but was moored by steel guy lines when driving the bucket system. With twin ten- by twelve-inch cylinders and cast steel gearing, the engine was linked to two drums—one thirty-three inches for hoisting and one fifty-four inches for hauling. The bucket was equipped with special digging teeth and was capable of carrying a ton of material per load, sixty loads per hour. It loaded and dumped itself automatically, its actions controlled by levers at the engine.[4]

Hanging teeth down from a carriage, the empty Knight bucket was run out on the cable and dropped to the ground. By carefully calculated braking and cable manipulation techniques, the bucket was then loaded, hoisted, and conveyed to the main tower. There it emptied its contents into the large built-in hopper, to be washed over a flat grizzly—a coarse screening device—separating rocks and boulders, then down a 200-foot sluiceway set on a five-degree grade, with water carried in a twelve-inch pipe from Granite Creek higher up. Caught behind riffles, the gold was periodically removed with quicksilver when the water was turned off for clean-up. The tower was located in a pit dug previously, and each year tailings were dumped in the previous year's excavation. When an area was fully worked, the entire plant was moved forward to begin its operations anew. Five men were needed to operate the cableway all the time: leverman, fireman, signalman, hopperman, and rigger. When the sluice was being run, which was only part of the time, pipeman and two assistants also had to be on board. And when the cableway worked full steam, as it did in the summer of 1898 and 1899, three shifts were employed. According to contemporary accounts, it took nearly thirty men to keep the cableway system operating on a twenty-three or twenty-four hour day.[5] Moving the apparatus was difficult and could be dangerous; occasionally, a workman was injured by a cable carrier or some other chance accident. And once a high wind blew over the A-frame—together with engine, boiler, and other equipment—causing "quite a loss."[6]

At first, the Lidgerwood equipment was apparently too light, and a heavier tower and machinery were installed. This "ingenious device, developed for us by the Lidgerwood company," as

Philip Moore called it, worked reasonably well on the relatively high grade gravel on the upper part of the German Bar ground, at least for the first year. Mining ground that had already been worked over "by the 'shovel in and shovel out' method" of hand placer mining, the equipment yielding by Moore's recollection thirty-seven cents per cubic yard at a cost of about twenty-four cents.[7]

But these profits were not obtained in the Conrey Ranch lands proper, and the cableway had limited capacity. Experts and onlookers did not agree as to how much gravel might be moved on a good day: estimates ranged from around 500 cubic yards in a twenty-four-hour period to a bucket record of 196 a shift, or 882 cubic yards in late May 1898. But the same company records indicate that the total output for May and June of that year was 15,596 buckets or 90,800 cubic yards, which yielded $3,650 in bullion at an operating expense of $4,500, an expense of 11.85 cents per yard.[8] For the 1897 season, the cableway took out $19,433.32 in gold at an average yield of 44.55 cents per cubic yard, according to the company, with costs at 31.46 cents a yard. In 1898, the company took out $28,904.51, an average of 36.52 cents per cubic yard at a cost of 22.6 cents, figures that seem suspiciously skewed. Output for May, June, and July 1898 showed an average return of 20.4 cents per cubic year at a cost of 31.56 cents.[9] Operations in the spring of 1900, when the cableway was still running, indicate that in April it excavated 28,792 cubic yards for $3,627.04, for a yield of 12.6 cents a cubic yard at a cost of 13.3 cents. But observers noted that accurate data on working costs was lacking for the life-span of the equipment and that the cost was probably close to 25 or even 27 cents a cubic yard.[10]

Clearly the Lidgerwood cableway apparatus was not cost efficient, although early prognoses were optimistic. According to the *Madisonian* after the close of the 1898 season, both the German Bar people and the kindred Conrey group were looking ahead, "having acquired a large piece of placer ground," a sure indication "that their new devices for mining flat placer ground are commercial successes."[11] Such conclusions were premature, to say the least. It was reported in the summer of 1899 that in the next year an even dozen dredges would be installed by the German Bar and Conrey Placer Mining Companies. These two concerns, "which are practically one," had commenced a series of experiments two years before. Some of their efforts had been successful and some had not, but "they have demonstrated the values contained in the farm lands at the mouth of Alder gulch, land that for thirty-five years had been used to raise cabbage and alfalfa upon."[12] To informed onlookers, the movable tower was like the early steam-operated traction dredges that had been used in Washington Gulch for dry mining, ". . . the first of their kind to be operated, and therefore largely experimental."[13] In the end, the Lidgerwood cableway failed to live up to its promises and in 1900 was abandoned. As one observer later commented: "The method proved too geometrical in its construction, and too costly in operation to admit of a reasonable profit."[14] Requiring nearly thirty men and handling only about 500 cubic yards in twenty-four hours, the unwieldy machine had little hope of success. "Like numerous other placer-mining plants of but slight mobility, resultant low capacity and high cost prevented successful operation," pronounced a dredge expert a few years after the system closed down.[15] Between 1897 and 1900, the Lidgerwood cableway worked some 380,000 cubic yards of gravel at Alder but at little or no profit. In subsequent years, the cableway was carried on the books showing a cost of $27,171.02.[16]

Overlapping the Lidgerwood experiment in Alder Gulch were efforts to adapt another modern innovation that had also been very much in evidence in the construction of the Chicago drainage canal—the Hayward steam bucket. At its plant on Church Street in New York City, the Hayward Company produced machinery for a wide variety of excavating and earth-moving operations: grading and digging for railroads; dredging and deepening harbors and canals; building sewers and sinking wells; submerging cylinders and caissons in bridge pier construction; digging ore and crushing rock; dredging phosphate; and, as the firm's final category, "placer mining and cellar digging."[17]

Hayward machines were well-known and had been used in California for mining as early as 1886. Some were of the clamshell variety, with a two-part grab apparatus hung on a cable at the end of a boom. Others, including those used by the Conrey group, were labeled "orange peel" buckets because they were built of three or four top-hinged curved triangular blades that closed to form a semi-spherical bowl to hold the excavated material. When open, the Hayward Company argued, the blades resembled sharp spades and were so coordinated with the power wheel that little dirt was lost as they closed. Although manufactured in New York, this equipment was actually an American version of the Priestman "grab," a British invention widely used in various parts of the world.[18]

While records are not clear, even before the cableway had been abandoned the Conrey people had invested in at least one Hayward "orange peel" apparatus—and probably two—to work the area just below the mouth of Alder Gulch. The new four-bladed bucket (or buckets) of two-cubic-yard capacity hung from a boom hooked to an A-frame on a large car that also housed its steam power plant. The excavating unit was mounted on runners, which moved over timber ties set on the ground at right angles for mobility. As it dug, the "orange peel" dumped its gravel into a high hopper. The gravel then went to a washing apparatus, which apparently was initially also mounted on cars, although illustrations in the Hayward catalogue and in the mining periodicals show the hopper as part of a movable trestle holding a 200-foot sluice mounted on wheels. The machine—or probably two—worked only a few months of 1898 in the gulch, dug 7,930 cubic yards, and recovered $1,752 in bullion at an operating cost of $7,488. The total cost, carried on the Conrey books, was $33,160.76.[19]

Perhaps the "orange peel" bucket was designed as much for prospecting as for actual full-scale mining. At least the editor of the *Alder Gulch Times* so reported in 1899, when he noted that the four-pronged clam shell had prospected to bedrock, digging a pit 200 feet square. But shrewd mining men quickly saw its shortcomings. The "orange peel" bucket was too light and lacked the power to dig hard gravel at a reasonable cost. Not only did it possess "the many faults common to placer-mining machines mounted on cars, it was also incapable of excavating the tenacious gravel." Further, despite the manufacturer's claims to the contrary, it lost much of the sand and small gravel through the openings between the leaves of the bucket. It is understandable that the company abandoned the "orange peel" so quickly.[20]

The Conrey innovators continued to borrow and to adapt. If the cableway and the "orange peels" had been used in constructing the Chicago Drainage Canal, so had a basic steam-powered dredge, used for years for building canals and harbors in various parts of the world.[21] Meanwhile, Samuel S. Harper and his Lake Michigan backers had amazed the mining world with the success of the *Fielding L. Graves* on Grasshopper Creek near Bannack, Montana, and had directly or indirectly attracted four more bucket dredges to that stream. And in San Francisco, the Risdon Iron Works was translating Robert Postlethwaite's plans of tried and tested New Zealand machines into working dredges on California rivers. The Conrey people could not have been ignorant of either development. Certainly, the spectacular Grasshopper Creek enterprises were practically in their backyard.

So it is not surprising that in the summer of 1899 the *Alder Gulch Times* announced that the Conrey Placer Mining Company now owned three miles of excellent placer claims, was at work building its own dredge, and was leasing ground to the Chicago Mining and Development Company to be worked by the *Maggie Gibson*, a large double-lift elevator dredge brought over from Bannack. "Double-lift" meant that the gravel was raised by the buckets and dumped into a hopper; the finer material was screened out and then lifted by a pump to the sluice or saving-tables. The same editor predicted that Madison County was on the eve of "an era of dredging machines"; that the German Bar and Conrey companies—"really one concern"—planned to install twelve monster dredges, each costing nearly $100,000; and that they would work Alder Gulch for the next thirty years. The Conrey group had proceeded shrewdly, the newsman observed. Experiments conducted over the previous several years at first gave the impression, "probably by design," that results had been "another gigantic failure." Meanwhile, "satisfied that it had the world by the tail and on a down-hill pull," the company quietly acquired ground up and down the gulch "until it had nearly everything that was of any value." The *Maggie Gibson* was removed from Bannack "and installed [on Alder Gulch] with stories of success rivaling the lamp of Aladdin."[22] So ran the press reports.

How successful the *Maggie Gibson* may have been at first is debatable, but certainly it established the practicality of profitable dredging of placer ground—even ground that had been worked once before by primitive methods. Originally built as a suction dredge with a revolving cutter similar to those on harbor dredges and with a sixteen-inch centrifugal pump to carry gravel to sluice boxes located on scows, the *Maggie Gibson* had been launched under another name at Bannack in July 1896 by the Chicago Mining and Development Company.[23] But as it had elsewhere, the suction dredge idea proved to be both impractical and unworkable, and in the spring of 1897 the re-christened *Maggie Gibson* (after the "estimable better half" of Archie Gibson of the Hotel Bannack), reappeared as a double-lift elevator-type bucket dredge.[24] As such, the dredge operated for two seasons on Grasshopper Creek with varying degrees of success, then ran out of paying ground. It was then that it was dismantled, rebuilt with a new hull, and moved in pieces to Alder Gulch.[25]

The *Maggie* was an open-connected dredge with five-cubic-foot buckets of about six hundred pounds each. Bucyrus-built, with four fifty-horsepower boilers, the *Maggie*'s steam engines burned from sixteen to twenty cords of wood per working day. The gravel it dug was elevated and dumped from the buckets into a revolving grizzly or trommel, where rocks and fragments over four inches in diameter were discharged over the side of the dredge. Passing to a sump in the hull, the finer material was then raised seventeen feet by a twelve-inch pump and was washed through a sluice about eighty feet long, sustained at the end by an auxiliary scow. It was from the sluice, fitted with riffles and charged with quicksilver, that the gold was recovered. Like most early dredges, the *Maggie Gibson* had plenty of defects: the trussed I-beam digging ladder was too light; the upper tumbler sprocket wheel was weak; and the 18- by 48-inch, 30 foot-long wooden spuds fitted their casings poorly and often broke (in time, they were replaced by a system of head-lines and mooring cables). These and "other features of parallel weaknesses" caused numerous breakdowns, yet the *Maggie* came to be considered not only the first dredge to work Alder Gulch but also a highly successful one.[26]

Despite erroneous early reports that the Conrey group had purchased the *Maggie Gibson*, the Chicago Mining and Development Company's double-decked dredge ran for four and a half years on ground leased from Conrey. Conrey was paid a sliding scale of royalties that began with 12.5 per cent of gross output of gravel running fifteen cents of gold per cubic yard and going as high as 60 per cent for forty-five-cent ground. Overall, this was perhaps not a major part of Conrey's income, but the company did gain $45,158 from the leasing arrangement before the *Maggie* was taken out of commission in 1903.[27] At that time, Montanans considered it to be one of the most advanced dredges ever built.[28] More important than the income she brought in was her influence on the Conrey company's development and perfection of the sluice-type dredge, a progression somewhat independent of dredge evolution in California.[29]

Chapter Four

An Interlude of Steam

At the same time they were leasing ground to the owners of the *Maggie Gibson*, the Conrey associates were building a dredge of their own. Unlike most of the Bannack dredgers, whose machinery was manufactured pretty much as a single unit by such firms as Bucyrus of South Milwaukee, the first Conrey dredge was a hybrid built primarily by the Vulcan Iron Works of Toledo, a firm that had been building steam shovels and dippers for a quarter of a century. A number of other concerns had sought the contract, but H. G. Bixby agreed with Philip Moore early in 1899 that while a Chicago firm was offering a price about $3,000 lower than Vulcan, it would probably be too slow, as "the whole business is new to them."[1] Bixby also wrote Gordon McKay two weeks later that he was pleased to have the order for the dredge placed and that the Risdon Iron Works of San Francisco had applied considerable pressure: "The Risdon people are very anxious to get an order and have recently sent me a telegram of ninety-one words from San Francisco." He went on to write that one of his friends, Major McNeill, had recently told him that their Risdon dredge in California "was giving them no end of trouble, and they seemed to be constantly finding additional weak spots."[2] The eager Risdon spokesmen, "anxious to get in with us," generally figured about six thousand dollars a year for repairs and depreciation on a modern dredge, but he cited one of theirs working the Feather River in California with repair costs, including all supplies but not depreciation, at not over two thousand for the year. By that time, Risdon had already built thirteen gold dredges and had seven or eight more under construction, mainly for satisfied prior owners.[3]

In any event, early in 1899 the Conrey people awarded the dredge-building contract to the Vulcan Iron Works Company, a firm that had already provided some equipment at Alder Gulch, probably for the Lidgerwood cableway at German Bar.[4] Work began to complete the machine in a hurry. By May 1899, records showed that the Conrey group had paid out $5,666.67 to the Vulcan Iron Works, not including $1,040 in railroad freight and $236.96 for wagon haulage of machinery, and that $4,257.78 had already been expended on construction of a hull.[5]

Completed in the fall of 1899, the new dredge was dubbed the *Gordon*, after Gordon McKay, or later simply "No. 1." Built with a chain of forty-one open-connected buckets of five-cubic-foot capacity, the *Gordon* had drawn inspiration from the *Maggie Gibson*, but it failed to resemble the earlier dredge in most respects. In size, the two dredges were comparable: at 110 feet, the *Maggie*'s hull was 10 feet longer than the *Gordon*'s, which in turn was 44 feet wide as opposed to 35 feet for the *Maggie*. Both were sluice-type dredges, but the new *Gordon* was a single-lift machine, capable of digging 35 feet below water level, whereas the double-lift *Maggie* could excavate only to 29 feet. The *Gordon* had a 140-foot sluice lined with angle-iron riffles, of which 120 feet extended outside the hull with one end sustained on an auxiliary barge. Sphericle joint-socket coupling allowed the flume to be swung sideways to give a better distribution of the tailings. Originally, the sluice was divided into two parts by a longitudinal partition to allow continuous running of one half while the other was being cleaned up. When this proved to be ineffective, the double sluice was abandoned. The big differences in the two dredges were in weight and power. The *Gordon* bucket chain weighed more than twice as much as the 40,000 pounds of the *Maggie* chain, and the total horsepower of the *Gordon*'s three-boiler steam plant was 176 compared to 100 for the Bucyrus dredge.[6] Originally, the *Maggie* had used spuds for anchorage but changed to lines, which the *Gordon* had used from the beginning. Using an intricate battery of winches and a network of heavy cables from bow and stern and anchored on shore at four or more points by "deadmen" (usually heavy logs buried parallel to the hull), the head-line system saved time and made stacking easier than a spud-mooring arrangement; but it was less effective on hard ground. The *Gordon* had twice as much gold-saving area as the *Maggie*, and the average digging capacity per month was 41,360 against the *Maggie's* 27,000. In general, the *Gordon* was much heavier, more sophisticated, and more powerful. At the time, it was the best dredge known.[7]

Nonetheless, when the *Gordon* first went to work in the fall of 1899, it was anything but a success. The ground proved to be "of an unusually compact nature, and though the dredge was apparently built of the same strength as dredges operated successfully in other places, it was unable to dig the gravel without continual and excessive breakages."[8] Efforts to use a Hayward bucket in front of the dredge either did not materialize or proved to be unworkable, and breakdowns brought frequent and costly delays. In April 1900, for example, the *Gordon* actually ran only 178 hours and 20 minutes out of a possible 270 hours and 15 minutes, a 34 per cent loss due to stoppages of all kinds.[9] As a result, the dredge soon "went through many vicissitudes of alterations, reconstructions and general repairs, necessitated by the severe duties with which it was taxed and which fairly strained it to the breaking point."[10] Rebuilding proceeded off and on for the better part of two years and was an expensive proposition. Conrey directors accounts show substantial payments for work for the *Gordon* to Midvale Steel Company, the Fulton Iron Works, and Scullin-Gallagher Iron & Steel Company. Some of the boilers, engines, and pumps used in the work came from the unsuccessful Hayward orange peel machines. As of the end of December 1901, the company listed the *Gordon* as costing $50,409.94, with an additional $26,944.69 spent on "winter repairs" apart from spare parts and operations.[11] Subsequent balance sheets showed $80,598 in capital outlay for the *Gordon* and renovation and repair costs of $157,290.[12]

The *Gordon* was overhauled and strengthened. Buckets of 5 cubic feet were replaced by larger ones, measuring 7.5 cubic feet and weighing nearly twice as much as the originals.[13] After a season, the renovated dredge was still not satisfactory and more work was undertaken. Eventually, the *Gordon* "was practically rebuilt, all the weak points were thoroughly over-hauled and strengthened to a standard required by the ground and since that time it has operated with exceptional success," according to one observer in 1902.[14]

That statement was too optimistic. Trial and error dominated the renovation, and not every change was an improvement. In fact, the renovation itself sometimes need renovation. Stoppages and repairs in the field continued, even apart from the extensive (and expensive) winter modifications. Annual repair bills were considerable: $4,312.71 in 1901, $6,732.10 in 1902, $6,678.25 in 1903, and $13,430.93 in 1904.[15]

While the *Gordon* was being rebuilt, the Conrey company was also busy constructing a larger steam dredge, the No. 2. Julius Baier, who had succeeded Philip Moore as company manager, arrived from Boston in October 1900 to commence active planning. En route, he had stopped in St. Louis, where he had family, and had ordered materials for the new machine.[16] The thirty-six-year-old Baier had mining experience in Colorado, although how much in the still new field of gold dredging is not clear. Like the *Gordon*, the No. 2 was a mongrel, patched together from designs and ideas of many men, some in Montana, some elsewhere. The Chicago Bridge & Iron Company and the American Bridge Company provided some of the major parts and assemblies, but the Stanley Manufacturing Company of Boston was most important. It supplied all shop work at 15 per cent above working costs and made the drawings and designs, no doubt because of its trustee, Frank F. Stanley, a long-time business confidant of McKay's, who by early 1901 was also secretary-treasurer and a substantial stockholder in the Conrey Placer Mining Company. Again, however, there was no central, experienced dredge builder, such as Bucyrus or Marion.[17] To utilize the available expertise, the Conrey directors voted to keep Baier on the job as supervisor but to bring the Stanley Company's W. O. Hildreth to Alder Gulch to install and start up the machinery. They also agreed to hire two men with experience on the *Maggie Gibson* to put an extra digging engine on the No. 2 dredge, which was not yet completed.[18] Looking back a few years later, engineer Hennen Jennings descried the fact that so many "experts" had been involved in designing and planning the early steam dredges that "great harmony in opinion as to details was not possible." There was no point, he believed, in running through

> . . . all the alterations, repairs and mechanical struggles, difficulties, unpleasantness, etc. encountered. . . . It is a case of beginning with too high mechanical ideals, no guiding experience, excess experimentation and education piecemeal.

Not only was the company's early history reflected in heavy construction and repair expenditures, said Jennings, but "the scrap heap also bears testimony to the variety and the non-success of the machinery used."[19]

In any event, with the Stanley Manufacturing Company coordinating activity, the dredge building at Alder Gulch proceeded under Baier's supervision. A branch of the Northern Pacific Railroad had been completed to Alder late in 1901, with the Conrey Company earning part of the credit for getting it built. But the road was not finished in time for No. 2, whose heavy machinery came by rail to Twin Bridges and was then laboriously hauled by wagon the rest of the way. One piece of equipment required sixteen horses to pull it, with six men walking along to supervise; the journey was made at the rate of about eleven miles in three days.[20]

Building the No. 2 was a slow process. Even before the dredge was completed, Conrey spent $34,097.14 on "No. 2 Reconstruction," indicating perhaps a last-minute change of mind.[21] Finally, in the summer of 1902, the crew of the No. 2 fired up boilers "and everything started off like clockwork," according to a local editor, who proudly extolled the virtues of this new and improved prototype.[22] The No. 2, reported the same newsman a few months later in the waning hours of 1902, was still in the field and would continue to work "until compelled to close by stress of weather and may possibly run all winter."[23]

Early observers believed that the No. 2 dredge "contained many improvements and embodied the experience up to that time," but it was soon clear that it was inadequate for long-term heavy-duty operation.[24] Newspapers continued to report "extensive repairs" on both steam dredges from time to time, as their owners sought the proper combination to work the Alder Gulch gravels. But No. 2, while much improved over the *Gordon*, was not strong enough to handle digging under severe conditions.[25]

How did the two Conrey steam dredges compare? Both were single-lift sluice-type dredges using head-lines, not spuds, for anchorage. The hulls of both were roughly the same dimensions, although No. 2 had slightly greater draft than No. 1 because of heavier machinery on board. Both dredges saved the gold in two-riffled sluices, one on the main boat itself, the other supported by an auxiliary pontoon and attached by a socket-joint coupling to the dredge sluice to give flexibility and a more uniform disposition of tailings. On No. 2, the bucket-chain was comprised of forty-one buckets, each with a capacity of 10 cubic

Statistics of Steam Dredges on Conrey Properties,
Ruby, Montana

	Maggie Gibson	*Gordon*	Number 2
Date built	1895	1899	1901
Type	Double-lift sluice	Single-lift sluice	Single-lift sluice
Av. Mo. Yds. worked, cu. yds	27,000	41,360	52,520
Drive	Sprocket chain	Sprocket chain	Sprocket chain
Bucket capacity, in cubic feet	5	5-7.5	5-10
Number buckets in chain	35	41	41
Buckets dumped per minute	7	5-11	5-11
Gallons water per minute	4,000	6,500	8,100
Total gold-saving area, in square feet	320	656	692
Horsepower of trommel	5	6	8
Horsepower of pumps	40	79	147
Horsepower of winches	10	12	4
Miscellaneous horsepower	None	19	16
Total horsepower	100	176	248
Hull dimensions, ft.	110 × 35	100 × 44	110 × 44
Trommel length, ft.	14′	16′	15′2″
Number of men	13	15	15
Costs of dredge	$225,000	$225,000	$225,000
Digging depth, below water, in feet	29	35	40
Length of flume, ft.	103	120	120

Source: Jennings, *History of Dredging*, pp. 11-14; Truman D. Prier, "Electric Power in Gold Dredging," *The Electric Journal*, 13 (December 1916): 563.

yards as opposed to the 7.5-cubic-yard buckets of the renovated No. 1; the weight of No. 2's bucket-chain totaled 169,000 pounds, 44,000 pounds more than the No. 1. On the new dredge, the upper tumbler was larger and at 4,050 pounds was more than twice as heavy as the comparable mechanism on the old. There were three pumps on the No. 2 instead of two, apparently an innovation of Gordon McKay himself. The new dredge had a total horsepower of 248, while the No. 1 had 176, but only slightly more gold-saving table space.[26]

Even though it was reconstructed several times, the Conrey No. 1, or the *Old Gordon*, as it was sometimes called, would be abandoned within a few years. Its machinery was simply too light and inefficiently designed. Dredge No. 2, finished in 1902, contained a number of improvements that were the result of the company's experience, but it too lacked the strength to cope with severe digging in tenacious clays. Breakage was too frequent and too expensive. The cost of winter modification work in 1902-1903 ran $13,431.41; for 1904, the cost was $28,690.49. Informed observers later reached the conclusion that No. 2 had been "well-nigh worn out by the end of the season 1906."[27]

Whatever the shortcomings of the Conrey machines, local tub-thumpers showered the dredges with praise, although one bruised drummer whose horses ran away after being frightened by the *Gordon* no doubt had different comments. The dredges in Alder Gulch reminded Laurin resident S. P. Page "of the steamboats on the Mississippi in the early days."[28] In 1902, the Montana Bureau of Agriculture, Labor and Industry applauded "the magnificent gold winning wizards of the Conrey Placer Mining Company" and proudly listed dredge mining under the category of "New Industries," along with petroleum, a woolen mill, and the manufacturing of grindstones and macaroni.[29] By 1904 one homegrown editor considered dredging to be "one of Madison County's best industries," one worthy of being publicized by brochure at the upcoming World's Fair in St. Louis. The Conrey Placer Mining Company was already "a permanent institution," with "ground enough to keep its dredges at work for two or three generations." And when the Montana World's Fair Commission circulated its free pamphlet at the Louisiana Purchase Exposition later that year, dredging activity in Alder Gulch—"the richest placer gold deposit ever found in the history of the world"—was not ignored.[30]

Like the *Maggie Gibson* before them, the *Gordon* and the No. 2 dredges were steam-powered. Initially, steam may have been a natural source of power in an isolated area, although it soon became apparent that it had numerous disadvantages. To begin with, fuel was expensive. In 1900, wood cost four dollars a cord and was hard to get at that.[31] Coal became the preferred fuel, but it was scarce and had to be hauled from Alder or sometimes Laurin, if the roads were bad. Individual freighters contracted to deliver the coal to scows on the ponds; and then it had to be transferred to the dredges themselves, a laborious procedure at best.[32] As time passed, the coal supply became uncertain, partly because of transportation and partly because of labor disputes at the mines. As the demand for coal increased, the quality declined; and in bad weather, the physical task of coaling the dredges, which were in hard-to-reach positions, often became even more difficult. It soon became clear that a new source of power would be welcome.[33]

Another problem with using steam power was the use of muddy pond water in the boilers, a practice that was often both necessary and deleterious to the life of the expensive equipment. Moreover, boilers, filters, heaters, and other gear connected to steam engine operation occupied a great amount of space that might better have been used for expanded gold-saving equipment.[34]

Nathaniel Shaler and other officers of the Conrey company well realized the shortcomings of steam. They were aware that the operating costs of the steam dredges ran about 16.5 cents a cubic yard. And they know that the 6 million yards of richest gravel worked by the *Maggie Gibson*, the *Gordon*, and the No. 2 below the mouth of Alder Gulch averaged about 22 cents a cubic yard, leaving little for profit when the loss of capital caused by abandonment of the dredge was taken into consideration.[35]

As early as 1904 Conrey officials were actively pursuing electrical power as an alternative to steam and at one point prepared to generate power on their own from a central steam plant. Manager Julius Baier mentioned in February of that year that Conrey would probably build its own power plant near Alder, although the company was also talking with a potential local supplier in Madison Canyon.[36] In September, Shaler proposed building an electric dredge that would use power from the Bismarck-Nugget Gulch Company plant or from Conrey's own engine-driven generator to be built at Alder.[37] Some action was apparently taken on the second course, for in December of 1905 it was announced that material for the power plant was beginning to arrive and would be pushed, as weather permitted, to provide power for both the dredges and for Ruby.[38] The company sent engineers and a crew into the upper Madison basin to scout out possible power sites, perhaps a strategic move to enhance its bargaining power while it negotiated for the purchase of electricity from the Madison River Power Company, the forerunner of the Montana Power Company. Later, after the bargaining appeared to have broken down, an agreement was reached with the power company, and a line was built from lower Madison Canyon to a substation at Ruby.[39] In this way electricity was brought from a plant below Ennis in 1906 to the small town of Ruby a full five years before larger centers like Dillon and Sheridan were electrified.[40]

When Gordon McKay died in October 1903, the bulk of his estate went to Harvard University, "tied up so ingeniously and so intelligently as to insure a long and steady increase in the resources of the Graduate Schools of Applied Science." When the last payments were made in 1949, the total was over $15,766,000, "the largest single bequest" to Harvard to that time.[41] Included in the bequest to Harvard was McKay's controlling interest in the Conrey Placer Mining Company. Nathaniel Shaler who had been managing director of the company, kept a personal eye on operations in Montana and acted as liaison with the officials of Harvard University, the new leading stockholder. After McKay's death, he was elected president at an annual salary of $5,000; Philip Moore was named vice-president.[42]

Shaler was well aware of the revolution that dredging was generating. In 1901, for example, he had predicted that within three decades the world would be experiencing a glut of gold as a result. He guided the company for several years, visiting Ruby occasionally and doing much to shape policy; ironically, he even had to fight hard in Cambridge to keep the McKay funds from being diverted to MIT. Active to the end, the well-liked professor-geologist died in April 1906, and James J. Myers of Cambridge assumed the presidency of the company. The socially active Myers was "first and last a Harvard man," according to one who knew him well. A lawyer who practiced in Boston, he was a staunch Republican who had served a decade in the Massachusetts legislature, including four years as Speaker of the House of Representatives.[43]

A month before Shaler's death in 1906, another of the pioneering Conrey founders, Philip Moore, sold his interest to Hennen Jennings and his associates. Jennings assumed Moore's old role of director and vice-president, and when Shaler died he also became consulting engineer.[44] Jennings had studied with Shaler at Harvard and had never met McKay, but he had conducted personal business matters with both James Myers and Frank Stanley, Conrey's secretary. For over a quarter of a century, Jennings had built a solid reputation as a mining engineer in some of the most famous mines of his day—the New Almaden in California, the El Callao in Venezuela, and the Robinson on the Rand in South Africa, where he pioneered the introduction of the cyanide process. After his work in South Africa, he became president of the British Institute of Mining and Metallurgy, a singular distinction for an American. Personable, industrious, and dependable, Jennings came to be regarded as an effective innovator and as a man of sound advice. Jennings was recognized as one of the outstanding engineers of the turn-of-the-century global mineral industry, and when he joined the Conrey Placer Mining Company as an active, experienced consultant, a new period of serious appraisal and reassessment was inaugurated.[45]

Jennings pushed the electrification movement that Shaler had begun, and he was responsible for a comprehensive report detailing dredge operations at Ruby and pointing out their shortcomings. He believed that the Conrey Company had begun with the best type of dredge then known and had "advanced it far beyond anything found elsewhere." But circumstances had changed. The open-link chain bucket used on the *Maggie Gibson* and both Conrey dredges had soon become obsolete, but the company was slow to adjust. And "matters were not aided by having an excess of mechanical designers, divided between Montana and Massachusetts." The result was a lack of standardization of machinery, compounded by the fact that Nathaniel Shaler— "the master mind," according to Jennings, in conceiving and organizing the project—could rarely visit Montana and provide continuity. The Conrey dredges worked more difficult ground than most contemporary California dredges at higher operating costs, but it made no sense to Jennings that the No. 2 had such large buckets but lacked the power to run with them full.[46]

It was Jennings who observed that the yield of 22.5 cents per cubic yard was higher than the California average, but that the 16.5 cent per yard operating cost over eight years was considerably more than the 6- or 7-cent-per-yard cost often incurred in California. He admitted that the nature of the terrain and ground made digging in Alder Gulch more difficult than in most Yuba or Oroville gravels. The Conrey hybrid dredges left something to be desired mechanically, and perhaps they had been too loosely supervised from Boston. Although instrumental in getting the operation going, Shaler had been too busy to give much ongoing personal direction; and Jennings reported that the system of accounting was poor. But the real key to success was economy. Conrey must reduce its expenses, he argued, and electric power was the first step.[47]

A week after the company's directors considered Jenning's "Report," the engineer met with Harvard President Charles W. Eliot and Treasurer Charles Francis Adams II (nephew of his diplomat namesake) in the offices of the college on State Street in Boston to give them his recommendations.[48] A new electric-driven dredge was already under construction. Jennings advised closing out steam operations, and presumably President Eliot concurred. In November, the Conrey directorate accepted Jennings's "Report," urging that the two steam dredges be shut down in the near future.[49] The dredges would be written off at a value of about $225,000 each, a heavy capital investment, although Jennings pointed out that they had "served the useful purpose of giving valuable experience which aided in future successes."[50] But as former manager Adolph Wiseman confided to Jennings a few years later, "I have often wondered if the stockholders didn't do a great amount of unnecessary pioneering with the Alder gravels, or rather with the mechanism to handle them."[51]

The Board of Directors commended Jennings for his work as consulting engineer and gave him broad powers to draw up plans for new electric dredges. They directed him to find and hire a new manager, with the stipulation that the manager not be paid more than $6,000 per year.[52] To gain some guidelines for the future, Jennings arranged for one of the experts employed by Wendell P. Hammon, a pioneer California dredge owner, to examine Conrey's property and equipment, but floods north of Sacramento delayed his coming. To show the depth of their confidence, the directors also voted to send Jennings to Ruby, with the power, at his discretion, to repair or electrify either of the two steam dredges or to build two new ones and a new machine shop, involving expenditures up to $250,000.[53]

Jennings and George L. Holmes, the California dredge specialist, were at Ruby in September 1906.[54] Holmes reported at length to the Board on the equipment of the Conrey placer operation, and his remarks supported what the Conrey directors had already heard from Jennings himself. Holmes observed that the ground was no better and no worse than grounds in California. Bedrock and stratification were similar, and he could see no good reason for the higher operating costs. The California design was single-lift with a stacker, while at Ruby the early dredges were single-lift with extended sluices. At Ruby, the mechanism screened only a small percentage of very coarse materials and sluiced the rest astern; the California rigs gave a more thorough screening, then conveyed coarse tailings to the rear via their elevator systems. Holmes also noted that California dredge hulls contained 50 per cent more lumber than did the two Conrey steam dredges and were much tighter and tougher. At the same time, the Ruby dredges had higher upper tumblers, which meant longer digging ladders and a need for greater power. The open-connected buckets of the Conrey machines carried only half as much dirt as California close-linked buckets of the same size. Furthermore, steam was expensive and led to high repair costs. Holmes urged a number of changes, including strengthening of the hull of No. 1, remodeling the bucket line, changing to anchorage by spud, and converting to electricity—all at a cost of at least $45,798. In addition, the entire hull of No. 2 should be rebuilt, the machine shop expanded, and a retort house constructed.[55]

Holmes complained that the excessively high repair cost per cubic yard of both No. 1 and No. 2 outran the entire working cost per cubic yard of the average California dredge. Conrey had invested a million dollars, worked five years with two dredges, and built a third, but in so doing had expended the profits from the original machines. "This is not progression," Holmes argued; "it is decidedly retrogression." Dredges No. 1 and No. 2 were "peculiar in design and construction," he reported, and in most ways were not as good as the average commercial dredge. Their hulls were "weak in the extreme" and poorly constructed against vibration, and their machinery was too heavy. They lacked the continuity of superstructure that added strength to the California dredges—the continuous trusses from stern through bow. The Conrey group had taken no precautions to preserve timbers—not even paint—to prevent mildew or dry rot; many California dredges were painted inside and out once a year. Such shortcomings and neglect in Montana, Holmes noted, resulted in rapid deterioration: fastenings loosened and hulls became warped and distorted. He recommended that No. 1 and No. 2 be rebuilt into one machine, electrified and designed along California lines.[56]

The running time of a dredge was the key to its success, Holmes acknowledged, and Conrey reports showed lost time comparing favorably with California figures; but he noted that the Ruby statistics were based on a twenty-three-hour day, not twenty-four hours, as in California. He recommended that the Conrey operations achieve better systemization. Holmes also believed that the acting manager and the superintendent at Ruby had provincial ideas. They had "been left to work out their own salvation" and knew nothing of outside expertise. What they needed, said Holmes, was to get out and observe dredging techniques near Folsom or Marysville. He advised:

> . . . California has progressed in the dredging art more rapidly than Ruby, not because of superior natural conditions, easier dredging or matters of that sort, but because of the larger number of individuals operating in a given field and the exchange of ideas and experience amongst the operators.[57]

Jennings agreed that what the company needed was the help of a California specialist like Hammon to oversee the redesign and rebuilding of the Conrey dredges.[58] Shortly before Christmas 1906, Conrey No. 1 was shut down for repairs and Superintendent Charles Kammerer left for California to inspect gold dredges there. Later that winter, Adolph Wiseman, then acting manager, also traveled to California to observe dredging techniques and machinery.[59]

In the summer of 1907, it was reported that both No. 1 and No. 2 would be changed over to stacker and table-types, equipped with spuds, and converted to electricity. But No. 1 was nearly eight years old and its hull was "in dubious condition," and No. 2 hull was badly warped. It was suggested that neither of these hulls be used and that entirely new dredges be built. Observers were impressed with the different approach here and with the amount of information generously doled out to visitors. "The dredging industry owes much to Montana," wrote John P. Hutchins, one of the country's leading experts. "Formerly, operators at Bannack solved many problems and now those at Ruby, in a most admirable spirit, are continuing the good work."[60]

By 1907, steam as a source of power was on its way out. Coal cost about $4.90 a ton, and each dredge consumed roughly fifteen tons per day. Soon electricity would be available at a penny per kilowatt hour, and Conrey's new No. 3 dredge would clearly demonstrate the superiority of electricity over steam—at least over the Ruby steam dredges, which would soon be discarded.

Chapter Five

The Triumph of Electricity

In keeping with long-range plans to develop its properties and to modernize and electrify its fleet of dredges, Conrey officers and shareholders incorporated the Poor Farm Placer Mining Company in the summer of 1904. With main offices in Virginia City, the new firm had a nominal capital of $500,000 in $5,000 unassessable shares of $100 par value each. Its board of directors was comprised of men who were all active in the Conrey Placer Mining Company: Nathaniel Shaler, Frank F. Stanley, Henry M. Bigelow, James J. Myers, and Philip N. Moore. All shares were assigned: Bigelow held 1,500; Shaler and Moore had 1,000 each; Lillian W. G. Farrington had 500; and Stanley, Myers, and George Gilbert—all executors of the Gordon McKay will—collectively held 1,000.[1]

The purpose of the new company was to take over from Harry Bixby the 480-acre Collins Ranch and the so-called Poor Farm property. By late 1898, Bixby had acquired title to the Collins Ranch and 140 acres of what had been the county poor farm, holding it in trust per agreement. In 1901, the Madison County Commissioners had sold the company a block of ground north of the poor farm and the county road. With the poor farm came the "Manager's House" (the farm's central building), the county hospital (which became the company office), plus a barn and two cabins—all of which were later valued at $8,500.[2]

Shaler, acting as trustee after Bixby's death, officially deeded the poor farm and Collins land to the Poor Farm Placer Mining

Company.[3] Long before that, directors of the Conrey Placer Mining Company had concluded a contract by which their firm would work the Poor Farm Company's ground. At the end of March 1905, Conrey officials voted to take $28,000 of their sinking fund and to loan up to a total of $300,000 from future funds to pay for the new dredge now being built by the Poor Farm Placer Mining Company. Directors also agreed to borrow $50,000 from the estate of Gordon McKay to complete the first Poor Farm dredge—already called the No. 3 electric.[4] From its initial capital of $50,000, the Conrey Company had increased its capital to $200,000 in 1899, then to $300,000 at the end of 1900, and finally in September 1905 completed the expansion to $1 million begun three years earlier. Of this, in 1906 only $860,400 had been issued—$300,000 in common stock, $300,000 in First Preferred Shares bearing 6 per cent interest, and $260,400 in Second Preferred Shares paying 8 per cent interest, to be paid before dividends were set aside on common stock.[5] Eventually, No. 4 electric would also be built for the Poor Farm Company, but the two firms were so intertwined in terms of directors, stockholders, capital, management, and equipment that for all practical purposes they were one company. They were officially merged in 1915.

Plans for the first electric dredge began as early as November 1903, when the Conrey board of directors authorized Superintendent Julius Baier to order lumber for another hull as soon as designs were complete.[6] Before the end of 1904, lumber for the dredge was in place and fifteen carpenters were busy. By Christmas, some of the iron work had begun to arrive, and "many strangers were seen upon our streets," the newcomers working on the new dredge.[7] Earlier, the Conrey board had accepted Baier's suggestion that an extra bucket chain be built for the proposed new machine—the No. 3 dredge.[8] In March of 1905, Baier was instructed to accept the bid of the Marion Steam Shovel Company to furnish a ladder hoist, mooring winch, and side-swing winch mechanism for No. 3 at the price of $7,100.[9]

At that time, the Marion Steam Shovel Company of Marion, Ohio, was one of the nation's leading manufacturers of heavy digging machinery. Its sales orders indicate that other parts of the Conrey dredge were being built by different firms and reflect a mutual effort with Conrey people to apply experience from dredge construction in California, modifying brake equipment, repositioning motors, and using more cast steel instead of iron. Marion priced at least the ladder hoist equipment according to weight—ten cents a pound, for a total of 21,848 pounds. Conrey spokesmen agreed to this arrangement but asked for "the very best piece of mechanism you can turn out for the purpose" and left it to Marion's engineers to decide what was most "satisfactory and worksmanlike." Marion's activities were coordinated with those of the Westinghouse Electric and Manufacturing Company, which built the motors at East Pittsburgh.[10]

Before the planning and construction of the No. 3 was completed, Julius Baier died in May 1905, but work went on.[11] While the two steam dredges continued to operate around the clock, the new machine took shape. Early in 1906, when the No. 3 seemed to be nearly finished, observers saw in the booming town of Ruby a vision of a great future, a day "not far distant when there will be thousands of people residing there" and when in a decade or so at least a dozen giant dredges would be at work below the mouth of Alder Gulch.[12]

No. 3 went into commission in September 1906.[13] Based on experience with the two steam dredges, which had substantial hull distortion because of the weight of buckets and ladders, special "hog rods" were installed in the new boat to tie both sides together and prevent spreading at the bow.[14] But before No. 3 had dug a single bucket of gravel, Hennen Jennings was voicing his doubts even though Conrey had invested at least $198,691 in the new dredge. "The total cost in construction of this one dredge," Jennings wrote, "is probably a world record."[15] Like the old steam dredges, No. 3 was an unfortunate mongrel, he thought. There had not been enough planning to fit the dredge to the specific environment for which it was intended, in this case the new Poor Farm property. The No. 3 would dig to only forty-three feet below the waterline, not the fifty for which it was designed and not enough to reach bedrock on the Poor Farm tract. It was large but not strong and would be costly to operate. Its open-link bucket chain was badly designed and poorly constructed, and it did not dump all the material. The boat used three times as much water per cubic yard as most California dredges. Its aggregate horsepower of 450 was too elaborate and expensive, and its five variable speed motors lacked efficiency when running at low speeds. Its switchboard was a luxury with too many instruments. And its undercurrent grizzly designed to capture fines otherwise lost was improperly constructed and missed most of the fine gold. A better dredge could have been had by contracting the whole thing with Marion or Bucyrus of South Milwaukee or the Boston Machine Shop in Oroville, California. To improve and strengthen No. 3 properly, Jennings estimated, would require and investment of roughly $50,000 more.[16]

In spite of Jennings's misgivings, No. 3 had a productive life. The first of Conrey's fleet of electric dredges, it retained a number of features of the old dredges and was the last of the sluice-type rigs. In addition to the long tail sluice, the No. 3 was of single-lift variety that used a mooring-line system of anchorage, and her thirteen-cubic foot buckets were originally open-connected. Although writers tended to exaggerate the bucket capacity, for its day No. 3 was unquestionably a monster. A machine of this sort, Thomas A. Rickard observed while construction was still underway, was "beyond the transporting capacity of all the mules in creation."[17]

No. 3's hull was 130 by 48 feet, with a draft of 6 feet. The flume was 135 feet long at the end of the boat and 66 inches wide, lined all the way with iron riffles. Her 43 buckets dumped 15 times per minute. With the digging chain run by a 150-horsepower motor, 10 horsepower motors linked to the specially built winch swung the dredge from side to side. The revolving screen or trommel was 18 feet long and 61 inches in diameter, a straight cylinder with belt drive from a train of gearing driven by a 10 horsepower motor. The oversized material from the screen passed out the end, then through rock chutes into the pond. Water for the sluiceway came from three large pumps that lifted about 12,500 gallons per minute—roughly 1,000 miner's inches. But this system could not pump enough water to sluice all the material dug, and the company was soon considering changing to a table recovery system that would require less water, less power, and obviate the need to frequently change riffles, in 1907 "a considerable cost item." The material passed from the upper sluice into the 135-foot tail flume that was 66 inches wide with a 7 per cent grade and lined with angle-iron riffles. From the sluice, material was discharged evenly via two under-current flumes, each 50 feet long, onto 360 square feet of riffled table space on the flume scow below. Cleanups came every ten days. Tailings were discharged in order to leave open water between them and virgin ground at the sides of the cut to prevent overlap and loss. No. 3 also had a boiler and steam piping and two large Hart & Cross hot-air furnaces for heating both the workers and the false-bottomed sluice. Several other unusual features were incorporated into the No. 3. The digging motor was located on a solid steel superstructure immediately behind the upper tumbler, the framing of which was connected directly to the A-frame, whence led the mooring lines. A special "aerial spud" was perfected that had been developed on the two steam dredges. This spud was a spindle supported by a ball-bearing casting atop the wooden A-frame on the back of the boat; it acted as a versatile locus for maneuvering mooring lines radiating out to anchors on shore.[18]

No. 3 used about 1.4 kilowatt hours of electricity per cubic yard dredged, although this amount varied depending on the soil and weather. In April 1907, in "less indurated" soil, the dredge dug 108,000 cubic yards in 599 hours, using only 1.25 kilowatt hours per cubic yard, but in January of the same year cold digging limited the total to 81,000 cubic yards in 539 hours and pushed the rate to 1.7 kilowatt-hours per cubic yard. Still, while the old steam No. 1, with its 7.5-foot buckets, was digging about 1,800 cubic yards a day and steam No. 2, with larger 10-foot buckets, was digging about 2,300 cubic yards, No. 3 electric was handling 3,300 cubic yards in the same number of hours. These yardage figures might not be considered large by California standards, as John P. Hutchins noted, but Alder Gulch placers were at a disadvantage. Climate, more intractable gravels in some areas, plus a noticeable lack of coordination among the digging, screening, and sluicing phases all helped reduce the excavating capacity of the dredges at Ruby.[19] In short, Conrey placers management left something to be desired.

Compared with the steam dredges, however, No. 3 had "a steadier, surer pull, with much less stalling action, and but slight variation in the chain speed of hard digging. . . ." Buckets filled more readily, "and there is much less racking, jerking and jarring." And No. 3 had an innovation that was copied by builders elsewhere. Its 150-horsepower drive motor was geared to the upper tumbler shaft through two intermediate shafts to incorporate slippage friction, whereas sprocket or belt linkage had been the custom on earlier dredges. Yet, it was clear that the new electric dredge was using more power than it should and that its belting was not altogether satisfactory, although belts were preferred to gears because of the safety of their slippage when buckets hit obstructions.[20]

Early in 1908, despite a broken ladder chain that dropped the buckets in the pond, a common cold-weather accident, Hennen Jennings reported his pleasure that No. 3 was doing well. Expenses were down "to a reasonable basis," and the dredge was turning a profit of about $10,000 a month.[21] At the end of 1909, the No. 3 had worked thirty-four acres to an average depth of thirty-four feet in a period of twenty months. In early 1911, the dredge was still in operation, "giving fairly good service."[22]

In the summer of 1911, the No. 3 was remodeled to increase its efficiency. With a new extended ladder 15 feet longer than the original, the dredge could dig to 54 feet, but there was a change from 13 cubic foot, open-connected buckets to 9.5 foot, close-connected buckets. To handle the extra weight, two triangular pontoons were built under the front end of the water-logged hull on either side—all at a cost of $12,880.[23]

Despite its use of steam heat, dredge No. 3 normally shut down part of the winter because its long sluice froze. In 1909-1910, the dredge was closed down in early January and the crew sent out to cut wood, but it started up again as early as the second week in February. The schedule for the winter of 1912—which was particularly cold—was pretty much the same, and in 1916 the machine was shut down most of the winter months.[24]

In 1913, it became evident that No. 3's digging course had been poorly planned. The dredge showed a 25 per cent increase in yardage and a 10 per cent decrease in cost while working upstream, as opposed to running downstream, statistics that bore out manager Charles Kammerer's earlier contention that the No. 3 should have been built at the lower end of the property and worked up. In 1914, the Poor Farm ground was worked out and No. 3 moved to Conrey property, making a long trip across poor paying ground.[25] By this time, No. 3 was regarded as old, tired, and inefficient. The boat was so ancient that in 1915 the National Union Insurance Company canceled the insurance on it, and the Conrey people in Boston had to find an alternate insurer.[26] The dredge closed down for the last time on October 30, 1916. It was dry-docked, the digging chain removed, the power disconnected, and all doors and windows protected and locked. No. 3 became merely a building—without a watchman—for the storage of equipment while the company sought a buyer.[27]

The construction on No. 3 had signaled the beginning of the age of electricity in Alder Gulch. At the end of 1906, as a result of the careful study by Hennen Jennings and California dredge expert George Holmes, the company took another dramatic step when it decided to abandon the older steam dredges and to reconstruct or build new machines of the latest design. At one point, the company contemplated having Wendell P. Hammon, through the Yuba Construction Company, build two new dredges at cost because of Harvard's connection with the Conrey operation. But these plans were superseded by contracts with the Marion Steam Shovel Company in Ohio to build the machines patterned on "the latest and most successful" California dredges, "with such modifications and requirements as the experience of the Conrey Company justified, particularly in respect to the nature of the climate and local conditions."[28]

It was June 1907 when the Conrey Company ordered two new dredges from the Marion, Ohio, builders, to be "put up and in running order" on or about January 1, 1908. The actual delivery date depended on arrival of the General Electric and Westinghouse motors listed in the specifications. Designated No. 1 and No. 2, both dredges were similar except for a few dimensional discrepancies. Marion agreed to furnish the machinery complete, set up in Alder Gulch. The builder prepaid the freight on No. 1, but records indicate that Conrey picked up the shipping tab for No. 2. Prices in the sales orders are not complete, but the figure $98,310 is mentioned in the job description for the new No. 1; presumably this does not include all of the expense of hull construction. Later, both Hennen Jennings and the mining press gave the total cost of each dredge as about $120,000.[29]

It was the company's original intent to utilize the two older steam dredges as much as possible, but this proved to be impracticable. As early as March 1907 work had begun at Ruby to convert the two dredges to electricity, but the idea had to be abandoned. The old No. 1 was entirely discarded. Orders to the Marion Company indicate that electric No. 2 was to be a remodeling of steam dredge No. 2, but in the end very little of the old was carried over. Only a small part of the woodwork and the plate girder head and brace rods from the gantries of the old dredges were usable on the new.[30]

Marion moved ahead, with shipment of machinery for both dredges to Montana completed between August 1907 and early March 1908. In the meantime, the Conrey directors voted to borrow up to $100,000 from the McKay estate to help pay for the new machines. Repayment would be made from dividends, with the company giving notes at 7 per cent.[31] During the first week of October 1907, twelve men from the Marion plant, under supervision of Howard Brenner, arrived at Ruby, to complete the hulls and install the equipment. Apparently, the work went well, except for a delay in repairing a winch motor for No. 2, occasioned when the motor was damaged in a fall, and another delay because the instruments for communications had not been shipped.[32] Even so, No. 2 began work on about June 1, 1908, about a month later than No. 1.[33]

Both of the new dredges were smaller than No. 3, and both were considered the latest state-of-the-art, California-style dredges. In the order files, the Marion Company specifications for various parts—among them the main drive, ladder hoist, winch machinery, and stacker hoist—are labeled "Perry practice" or "Hammon & Perry," indicating that they were to be patterned after designs by Wendell Hammon and O. B. Perry, California dredge pioneers who had been responsible for many early improvements.[34]

Both dredges carried 7.5-cubic-foot buckets, but No. 2's digging ladder was 11 feet longer than that of No. 1 and could dig to 35 feet rather than 30. Buckets on both were close-connected, although apparently No. 2 began with a used ladder of open-connected, 10-foot buckets, and then switched over. The hull dimensions of No. 1 were 96 by 44 feet, with a 9-foot depth; No. 2 was slightly larger, at 102 by 44 by 9 feet. Both required a pond of 200 by 125 feet; the larger No. 3 needed 200 by 210 feet.[35]

Both were electrically driven, with a total of 380 horsepower distributed to their Westinghouse or General Electric motors, the largest of which—a 100-horsepower, 2,200-volt reversible, variable speed model—drove the bucket chain. No. 1, at least, was fitted with a system of speaking tubes for communication.[36]

Both of the new dredges followed the California traditions of stackers, spuds, and gold-saving tables, something of a departure for Conrey. Gone were the long tail sluices for discharging the rubble, replaced now by the more popular canvas and rubber-belt stacker. The stacker carried the larger debris and dumped it from some height at the back end, while the small material washed over the tables and emptied into sluices that discharged it into the pond at the stern. Instead of dropping boulders under the hull, the stackers deposited them in long regular heaps.[37] Where the earlier dredges had depended on anchor and head lines, the new ones were each equipped with two massive spuds, one of wood and one of steel. Those on No. 1 were two feet by three feet in diameter and forty-eight feet in length; No. 2's spuds were six feet longer. The two spuds moved up and down in wooden casings at the stern that had been designed and strengthened to protect the hull. While the wooden spud was raised, the steel one anchored the dredge in place while digging, serving as a pivot from which the dredge was swung sideways at the bow. When the dredge moved ahead, the steel spud was winched up and the wooden one dropped as an axis. By alternating spuds, the dredge could be "stepped" or moved to any desired location. While operating, the machine was swung sideways in a cutting arc by two lines fastened to either side of the lower ladder, then to a sheave on the bank, back again to each side of the dredge, and to a motor winch on deck, each in turn pulled or released to pull the boat one way or another. Not only was this improvement over the sluice box and head-line system more efficient, but it was also less expensive to operate and it could handle all types of gravel without trouble.[38]

On both No. 1 and No. 2, the buckets dumped the dredged material into a hopper; then washed by powerful pumps it passed into the trommel or grizzly, a thirty-five-foot-long tapered cylinder ranging from fifty-one to seventy-two inches in diameter. While the oversized stuff went through and onto a conveyer belt to be dumped over the stern by the stacker, the "fines" passed through the trommel's three-eighths inch and one-half inch perforations to pans and a battery of twenty gold-saving tables, each thirty inches wide and eighteen feet long. All material then emptied into small sluices that carried the debris over the end of the boat. Once a week the gold was cleaned up, using mercury to create an amalgam.[39]

Though electrically driven, both dredges were equipped with steam-heating apparatus to keep both crews and equipment warm and to enable them to run in cold weather. At work, they maintained a constant oscillation that kept the water in motion and helped prevent freezing.[40]

No. 1 and No. 2 represented a breakthrough. They embodied the best of both California and Montana dredge experience to that date and were less experimental than previous Conrey machines. They were also more cost-effective, and they taught Conrey that quality was a vital factor regardless of expense and that larger size resulted in economy of operation.[41]

Having learned these lessons, the Conrey Company announced early in 1910 that it was ordering a fourth and even larger dredge from the Marion Steam Shovel Company to be built just below the No. 3 dredge to work the ground between Ruby and the town of Alder. No. 1 was working farthest up the Gulch. No. 2 was working about a mile above Ruby and, according to a visitor a few months later, was plowing through a field from which clover had just been cut, "while a few feet in front of the dredge was growing one of the finest oat crops in the state."[42]

Hennen Jennings and Superintendent Charles Kammerer traveled together to Ohio to order what Jennings predicted would be "the largest & I trust as perfect & strong a dredge as ever has been built." Construction was to begin on April 15 with an estimated one hundred men on the job.[43] By an agreement of February 18, 1910, in the name of the Poor Farm Placer Mining Company, the Conrey people called for the building of a fifteen-cubic-foot dredge, the shipment of parts to be made "as soon as possible, as Dredge is to be completed and ready to run by November 1, 1910." The terms were "the usual," and a $50,000 check accompanied the contract. As before, Marion would supervise construction, keep its experts on the scene for thirty days after start-up, replace free any defective parts during that period, and provide the Poor Farm Company with a set of blueprints "for their exclusive use and not to be copied or examined by outside parties." For each day the dredge was ready to run before November 1, 1910, the Marion Company would receive a premium of fifty dollars, and for each day after that date a fifty dollar penalty would be charged; but neither premium nor penalty was to run more than thirty days.[44]

For its part, the Poor Farm Company agreed to furnish 2,200-volt current for use during construction, with a line available near the pit where the dredge was to be built. It also agreed to deliver lumber, machinery, and all materials to the site as soon as it arrived at Alder by rail. Orders called for shipments in sequence, with bolts, rods, pins, plates for the framing and bottom of the boat to arrive first; followed by foundation casting, brackets for the main drive, ladder hanger shaft, and gantry materials; then stacker, revolving screen, pumps, electrical fittings, and upper tumblers; then the main winch, ladder hoist, operating levers, ladder hoist, operating levers, ladder, and cables; next the lower tumbler and bearings and spuds; with the final shipment of digging buckets—all at intervals of about two weeks. Orders specified the Poor Farm Company's request that machinery be shipped in flat cars, not gondolas or boxcars, because they had no crane for unloading the heavy stuff.[45] Notations at Marion indicate that all equipment and materials were shipped by October 8, 1910, on eighty-five railroad cars. (Conrey spokesmen had earlier estimated that sixty-eight cars would be required.) It was a remarkable performance by the Marion Steam Shovel Company; it had begun the job using a new set of drawings and patterns and yet was able to meet its contract obligations in a little over six months.[46] At an original cost of more than $260,000 and at the time considered the largest dredge in the world, the No. 4 was carefully planned. Specifications were blunt: "It is the understanding that it is the intention of The Marion Steam Shovel Company to make this the most complete and perfect dredge yet built."[47]

Work on No. 4 began immediately. During the first phase, workers dug an 18-foot-deep pit that measured 150 by 250 feet. Edward Johanson had competitively bid the contract at twenty cents a cubic yard (beating bids of fifty-five and sixty cents) and put an average of eleven teams a day on the job with Fresno scrapers. The work was completed in 19.5 days, and Johanson cleared $1,100 after all expenses. As the *Alder Gulch Times* editor put it, "A few more contracts like this and Ed will be on easy street."[48]

By early July, some fifty men were laboring on the new dredge, among them carpenters from Marion, Ohio, who would return home in the fall.[49] Harry Fies, a Marion engineer in charge of construction, took a bride, Madge Cox of Ruby, who would accompany him to Ohio when the project was completed.[50] George W. King, one of Marion's top officials, briefly looked in on the work at Ruby, although his chief reason seems to have been a visit to nearby Yellowstone National Park with his wife and daughter.[51]

As with No. 3, Howard H. Brenner of the Marion Company was in charge on the scene. While Hennen Jennings was enthusiastic about the big machine, he was disappointed that it was not on schedule in mid-April 1911, though Brenner expected the boat to be floated on about May 1.[52] When Jennings, Assistant Manager Truman Prier, and Superintendent M. E. Buck of the Madison River Power Company plant met near the plant at the end of April, No. 4 was finished and was expected to begin digging within a week or ten days. Jennings, who did not stay to see the start-up, was impressed with the completed machine, which struck him "as a splendid piece of work, and I doubt if there has ever been a stronger or better gold dredge built." The engineer might soon have occasion to doubt that statement.[53]

Designed to cut its way through banks fifteen feet high and to dig fifty-three feet below water level, No. 4 was a monster. The hull was 150 by 58 by 13 feet, with a 5-foot deck overhang on each side. It was made of 550,000 feet of Oregon pine, "of the quality known as No. 1 Ship Stuff" that was "thoroughly caulked with three strands per seam of the best Navy oakum." The stern was built of 14-inch timbers, the her bow panels were of 8-inch stock, and the 4.5-inch side and bottom panels were covered with quarter-inch metal plates to protect against punctures.[54]

No. 4 was designed with two ten-ton traveling cranes located over the main drive or upper tumbler gearing, to move the full width of the boat. A three-ton electric hoist was situated on a jib crane for handling buckets being repaired or replaced or to lift large boulders, chunks of ice, or frozen ground from the bucket line. Each of the 80 15-cubic-foot buckets was of high carbon, open-hearth steel, specially heat treated, and weighing about 4,100 pounds, exclusive of the heavy manganese steel lips.

The plate girder-type ladder that supported the buckets was 116 feet long and, with its 21 steel rollers, weighed 112 tons.[55]

The other machinery for the operation was to scale. Fifty feet long and nine feet nine inches in diameter, the massive revolving screen and its drive weighed eighty-two tons. The total gold-saving area of 3,000 square feet—larger than that of Conrey No. 1 and No. 2 combined—consisted of a bank of fourteen gold-saving tables set at right angles on each side of the trommel and emptied into seven more tables (thirty-six inches wide) running fore and aft, eventually washing into outboard sluices, two on each side of the boat. The 130-foot-long tailing stacker was of solid plate girders decked over with steel plate; it carried a 14-inch rubber stacker belt and was canvas-covered, and its bottom was equipped with a steel flue for steam heating. The stacker was raised or lowered by a hoist located in the upper cabin, which was directly connected with a 36-horsepower variable speed motor. The heaviest parts shipped in one piece were the metal spuds, each eighty feet long and weighing forty-two tons. It was impossible to get wagons or trucks strong enough to haul the spuds, so steel rails and ties had to be laid to get them the mile or so from the Alder railroad station to the dredge site.[56]

The winching machinery for the No. 4 consisted of eight drums, four for the side lines, two for the spuds, and two for head lines or for general purpose. Wire rope hoist lines were 1.5 inches in diameter, bow swing lines and spud lines were 1.25 inches, and both bow and stern gantry lines were 2.5 inches. The gantries themselves were massive frames from which the cables ran for raising and lowering the digging ladder in the front and the spuds and stacker in the rear. The main drive operating the bucket line was powered by a 300-horsepower General Electric motor and came equipped with a friction clutch on its first reduction shaft to disengage the motor and act as a slipping shaft to prevent breakage. Ladder and bucket line were raised and lowered by a massive hoist on the lower deck inside the house, driven by a separate 200-horsepower, variable speed, reversible-type motor that permitted independent up-and-down action—a definite improvement over using the digging motor to also power the ladder hoist on contemporary dredges.[57]

Worthington centrifugal pumps provided water for the gold-saving tables and to wash away debris. All electrical equipment came from General Electric. In addition to the digging and the ladder hoist motors, a 150-horsepower motor drove the 14-inch-high pressure pump; a 100-horsepower motor ran the low-pressure pump; 75-horsepower motors drove the screen drive, sand pump, and stacker hoist; a 50-horsepower motor ran the winch and stacker drive; and a smaller 35-horsepower motor actuated the 4-inch pump. All motors were direct geared and used no belts. As originally planned and built, the No. 4's system included a total of 1,050 horsepower.[58]

In a pilot house patterned after No. 3's, the new boat carried sophisticated control facilities. Twenty-four operating levers were arranged so that one man could control all movements of the dredge. A five-panel marble switchboard equipped with complex instruments governed a separate line (in conduit) to each motor, with all lines equipped with automatic circuit breakers. Like the other three Conrey electric dredges, No. 4 was hooked up to the substation, where the 46,000 volts from the Madison River plant was stepped down to 2,200. With transformers on shore, such high voltage for all motors meant less weight, a lower potential for fire, and a smaller loss of power, although it did pose a greater danger to workmen. Each dredge had its own automatic switch and meter, and each was part of a private telephone system linking the dredges with the offices and homes of the dredgemaster and other important employees.[59]

In truth, it did seem that the monster 2,200-ton No. 4 was the technological marvel of the 1911 mining world. "In designing this dredge," wrote the editor of *The Engineering and Mining Journal*, "no expense was spared to make it the most complete and perfect dredge yet constructed."[60]

When it was finished, No. 4 had a capacity to work 300,000 cubic yards per month, equal to the other three Conrey dredges combined. Because of this impressive statistic, it was anticipated that the company's output would be doubled, but output fell short of expectations.[61] The trommel, bucket line, and digging motor all proved to be inadequate and almost immediately had to be replaced or greatly strengthened. The faulty design and insufficient material in the defective washing screen was not detected during the thirty-day trial run. Only after much lost time and expenditure of some $20,000 in repairs was the screen replaced in 1912 with a new one valued at $27,500, which Marion provided practically at cost. Even so, this pushed the price of the dredge to $296,206.04. The replacement screen weighed nearly one hundred tons, with the entire screening surface made

Statistics of Electric Dredges, Conrey Placer Mining Company

	Number 1	Number 2	Number 3	Number 4
Date built	1908	1908	1906	1911
Type	Single lift table, stacker	Single lift table, stacker	Single lift table, sluice	Single lift table, stacker
Av. Mo. Yds. Worked, cu. yds.	96,270	62,709	82,415	300,000
Drive	belt	belt	gear	gear
Bucket capacity, cu. ft.	7.5	7.5	9.5	16 to 17
No. buckets in chain	60	80	80	80
Buckets dumped per minute	16	15	15	18 to 20
Gallons water per minute	6,700	6,700	12,500	12,000
Total gold-saving area, sq. ft.	1,231	1,231	1,264	3,000
Horsepower of bucket chain	100	100	150	550
Horsepower of trommel	30	30	10	100
Horsepower of pumps	160	160	175	285
Horsepower of winches	20	20	115	250
Miscellaneous horsepower	70	70	10	50
Total horsepower	380	380	460	1,235
Hull dimensions, ft.	44 x 96	44 x 102	48 x 130	58 x 150
Trommel length, ft.	35	35	18	48, 8 in.
Type of anchorage	spud	spud	lines	spud
Number of men	8	8	10	10
Cost of dredge	$108,000	$128,000	$198,000	$296,000
Digging depth, below water, ft.	30	35	45	55

Source: Jennings, *History of Dredging*, pp. 11-14; Prier, "Electric Power in Gold Dredging," *The Electric Journal*, 13 (December 1916): 563.

of one-inch perforated manganese steel plates, with agitating bars cast on them, and believed to be the strongest in use anywhere in the world. The bucket line and ladder were too weak for the job at hand and had to be lengthened and reinforced. In the process, Marion provided twenty-seven new buckets and $5,000 in cash. At a cost of $6,500, the 300-horsepower General Electric digging motor was replaced by a 550-horsepower Westinghouse model, with a special liquid rheostat switch control. The GE motor was kept in reserve and was utilized again in 1918, when the more powerful motor broke down. But these renovations and the overhauling and adjustments that went with them tied up the dredge during much of 1912 and the winter and spring of 1913.[62]

By early 1914, No. 4 was in shape. Hennen Jennings was finally convinced that there was "not a stronger or more perfect gold dredge in existence." No. 4 had handled 281,000 cubic yards in a single month, and more was now expected. In 1914, the new boat ran an average monthly tonnage of about 325,000 cubic yards, while all of Conrey's four dredges combined worked 6,300,000 cubic yards "and furnished the bulk of the placer gold in Montana."[63] The Age of Electricity was in full swing, and dredge technology in Alder Gulch was on the same course as industrial manufacturing across the nation.

Chapter Six

Navigating Alder Gulch

Delay was the nemesis of any dredge operator. Successful dredging required constant operation, day and night, and heavy capital investment could not be allowed to remain idle. Time lost was gold lost. In northern areas like Montana, the cold months constituted an additional limiting factor, although not to the same extent as that faced in the permafrost regions of Alaska and the Yukon. But the byword of dredging everywhere was efficiency. To work around the clock and the calendar and to obtain the greatest possible output in a given time was the goal and aspiration of Conrey and all other dredging operations.

When every working moment counted, replacement parts and shop facilities were essential, especially in a remote region like Montana. As early as 1902, the Conrey Company announced that it was contemplating a 30- by 100-foot machine shop to be equipped "in an up-to-date manner throughout." Built above the Conrey house at Ruby, the new structure would be impressive: it would have a steam hammer, drills, shaper, hydraulic press, metal lathes, and other sophisticated tools with which to thoroughly overhaul and remodel the existing dredges. Conrey directors also voted to stock a supply of spare parts.[1]

The original shop was destroyed by fire in the spring of 1908 but was replaced the following year by a greatly expanded machine and blacksmith works in a building 75 feet wide and 225 feet long. Among other equipment, all powered by a 50-horsepower electric motor, the new works boasted two steel planers, a special double-spindle boring mill, three lathes (42-inch, 24-inch, and 18-inch) for turning steel shafting, a 4-inch radial compressed air riveter for riveting buckets, and a 1,600-pound steam hammer for forging bucket pins and other heavy items. With eight skilled mechanics at work, the shop carried a large stock of spare parts and could manage almost any kind of dredge repair and machine work.[2]

In 1913, Hennen Jennings told W. W. Mein, another mining man, that the company had invested over fifty thousand dollars in its machine shops at Ruby, carried an additional forty or fifty thousand dollars worth of spare parts, and maintained "a generous general-store supply."[3] Overall operating costs indicate that approximately 30 per cent was spent on repair work, a function that Jennings believed made the machine shop, with its elaborate equipment and force of trained mechanics and machinists, absolutely vital. Workers repaired parts as large as twenty-four-ton tumblers, rehabilitated machinery from sunken dredges, made a steel truss from irons of a screen discarded from No. 4 dredge to cure the warping and settling of the hull of electric No. 2, and re-lipped and bushed bucket chains.[4]

Bucket lips had to be replaced every few months, depending on the type of ground the dredge worked. When all four dredges were working, the process was ongoing and honed to a fine technique. As Conrey's business manager described it in 1917, fifteen worn buckets on No. 4 were replaced temporarily with old carbon steel types—a nine-hour job—and hauled to the shop by traction engine. There, the shop crew could remove the old lip and rivet a new one on three buckets in nine hours; usually an extra shift was put on at night to speed up the work. Re-lipped buckets were replaced, and the next fifteen were brought into the shop until the entire line had been refurbished.[5]

The second machine shop and all of its machinery were destroyed on June 15, 1915, "a total loss from fire," with an estimated $125,000 going up in flames. Manager Charles Kammerer announced that the shop would be rebuilt, but the new shop would not be as extensive as the old one.[6] According to business supervisor George Edmunds several years later, when he compiled an inventory of the company's physical property, the new corrugated-iron machine shop cost only about $8,000, exclusive of machinery. Edmunds's records in early 1918 showed that the machine shop was insured for $18,000, of which $9,000 was for equipment and $5,000 for stock. Obviously, at a time when the Conrey operation was winding down, its shop investments had been no longer what they were in the days of four working dredges, but there were still eight men employed in that part of the operation in the summer of 1917.[7]

Weather was always a factor in operating and repairing the dredges. During the winter, extreme cold and stormy conditions determined the length of the dredging season. Although Montana conditions were not like those in Alaska, where permafrost posed serious problems for dredging and eventually necessitated the development of special cold-water thawing techniques, the winter cold did curtail operations. It was reported that "the Montana climate is very likely the severest where gold dredging is thus carried on during the winter."[8] Normally, dredge workers could expect to be unemployed for three or four months due to the cold.[9] Two days after Christmas 1902, for example, the *Alder Gulch Times* reported that the *Gordon* had closed down for the winter, but that the new No. 2 would continue to work "until compelled to close by stress of weather and may possibly run all winter."[10] Apparently, No. 2 shut down soon also, and the *Times* editor could report in February that the Conrey steam dredges would resume operation in late March.[11]

Over the years, using techniques developed by experience, Conrey managers learned to operate year-round when winters were moderate, as was the case in 1904 when the two steam dredges kept going through the coldest months. The temperature reached from six to seventeen degrees below zero only occasionally and for brief periods; most of the time the thermometer registered from zero degrees to freezing. On the average, the frost was only about two feet thick, and the Conrey people found that their buckets could handle frozen ground down to a foot or fifteen inches. But when the frost was deeper, it had to be broken in advance of the boat by small charges of dynamite—"giant powder"—set off in holes for every ten square feet or so. The 125-foot-long sheet-iron sluice was more readily affected by temperature change than the wooden sluices, and froze more easily, but by keeping the water moving and turning a steam hose against its bottom it could be kept free. The first freezing problem encountered was overcome by building a light wooden casing—an "overcoat"—to surround the bottom and sides of the sluice and warming the enclosed air space with a steam pipe, although the big No. 3 electric never completely solved all of its cold weather problems.[12]

The Conrey people also learned the technique of spreading water by means of low coffer dams in the fall and flooding the ground to be worked. This allowed the ground to freeze and provided a blanket that slowed the penetration of frost to a deeper level. They learned to warm the ponds with condensation water and to melt the ice topside with steam. They learned to attack the ice physically, chopping it laboriously from sides of the hull, digging ladder and chutes. And they learned to manipulate the dredges to get a maximum amount of sunlight on the parts most likely to freeze. If operating constantly, the dredges set up slight currents from the digging ladder, which kept the pond from freezing. Amalgamation of the gold was slower when it was cold and required more quicksilver, but it seemed to be just as thorough. In fact, records of steam dredges No. 1 and No. 2 in 1905 and 1906 show little correlation between low temperatures and either higher cost or lower yield. During the winter of 1906-1907, when the mercury dipped to thirty below zero Fahrenheit, Conrey No. 3 kept working and the pond water remained just above freezing, even though ice on the hull increased the draft of the dredge by 25 per cent. The ladder rollers were ice-covered and the buckets slid awkwardly. Frozen material accumulated in the hopper; small pipes and valves burst, and outboard pump suctions froze if the dredge shut down. But No. 3 ran 72 per cent of the time in January 1907, and only 5 per cent of the time lost was due to excessive cold. Dredge experts cited Conrey dredges as "the first to run throughout the winter in such a cold climate." With their modified design and maneuvering and with protected stackers and boiler heating, dredging could continue anywhere in the United States after all other outdoor mining had stopped, according to engineer John Power Hutchins in 1907.[13]

During exceptionally cold weather, ice formation on parts of the dredges might cause delay or breakage. In 1909, the weight of ice on No. 1 electric caused the tackle carrying the stacker to break, dropping the stacker on the tailings.[14] In December 1918, the same dredge had more ice troubles when water froze in and on the back of the buckets. Because the dredge was working in fine sand, there was not much coarse gravel to wear off the accumulation of ice.[15] Occasionally, bad weather might precipitate a freak accident, as it did in August 1915, when lightening struck No. 3. Fortunately, the only damage was a burned out motor.[16]

Other seasons presented different problems in Alder Gulch. When is was unusually dry, there might not be enough water to float the dredges or to hydraulic ten or fifteen feet of top material from high banks. The *Gordon* lost four hours while stuck on a rockpile one day in mid-April 1906, and a scarcity of water shut down steam dredge No. 2 seven times in the autumn of the same year, with delays ranging from fifty minutes to ten and a half hours. When water was low, it was often necessary to dig bedrock in order to maintain flotation, a process that not only wore down bucket lips quickly but that also took three times as long as digging gravel because the gummy bedrock material stuck to the buckets and only two or three inches could be dug at a time. "It is more of a scraping process than a digging process," one expert reported, "because if too much is gotten into the buckets at one time, it is almost impossible to clear the buckets."[17] The drought of the summer of 1918 ruined many Montana crops and reduced the need for hired hands, thus easing the labor shortage a little, but it made dredging difficult—especially if the boats were on "drifted ground" where it was hard enough to keep water in the pond. Big No. 4 lost a good deal of time that summer because of low water. In order to keep a water supply, No. 4's crews had to use the overburden to cement up tailing piles parallel to its cut. When it got too near its own holding dam, the dredge had to stop and gradually adjust the water. In the meantime, crews took the opportunity to put on a new stacker belt and pulley, but they could not install the new upper tumbler because the dredge could not be swung into the proper position.[18]

The pulse of a dredge was its lost time record. Most dredges kept logs of every minute lost on the daily shift report and summarized running time and gave detailed analyses of causes of stoppages. Such documents were often accurate barometers of profit and loss and predictions of corporate health. These historical records show that Conrey's earlier steam dredges lost a good deal more time than the later electric ones and that the company's concern with prevention and quick repair of breakdowns paid dividends, both figuratively and literally.

There were many reasons for lost time, some of them perfectly routine, expected, and unconnected with accidents or broken parts. Moving the dredge ahead was an intricate maneuver that reduced operating time considerably. On the steam dredges and on No. 3, this involved manipulating head-lines and side-lines to advance for the new cut. On electrics No. 1, No. 2, and No. 4, "stepping ahead" was achieved by loosening one set of side-lines, tightening the opposite ones, and using one spud as a pivot. With the bucket ladder and the digging spud both raised, the second spud was dropped and functioned as a fulcrum for swinging the hull at an angle of forty degrees or so. After that, the first spud was lowered, the second one elevated and the operation repeated, until the dredge had moved the five to ten feet forward necessary for a new arced cut, after which the digging spud and the bucket line could be dropped again, the side-lines reset, and dredging recommenced. While the dredge was "stepping ahead," the crew usually took the opportunity to lubricate the lower tumbler, which ran under water most of the time and was subject to harsh abrasion. In 1906, the *Gordon* lost from two and a half to five hours a week, and No. 2's moving delays ranged between two and a half and seven and a half hours during a comparable period. Changing swing-line settings took from a quarter of an hour to four hours once or twice a month. Cleanup of the gold took from four to six hours every two weeks, but on exceptional ground it might be done every week. Handling boulders was an accepted part of dredging, but it was a sporadic business that might cost from ten minutes to a half an hour every week or two. Choked hoppers happened every few weeks when rocks or debris dumped by the buckets failed to clear the hopper into the rotary screen and might take from fifteen minutes to an hour to clear.[19]

Regular repairs, such as re-lipping buckets or replacing the chain, could usually be anticipated in advance; but many stoppages were unexpected and resulted from breakage. Belts and sprocket chains broke or had to be adjusted, absorbing from ten minutes to three or four hours every two or three weeks. Both of the early steam dredges had constant trouble with their digging engines—trouble which absorbed from thirty minutes to as much as six hours every few weeks. Boiler repairs were common: No. 2's was worked on three times in the fall of 1906 and lost thirty-five hours and forty-five minutes as a result.[20]

A major cause of early stoppage was the digging chains. The chains seemed to have been one of the weakest parts of the two early steam dredges, according to Hennen Jennings, who noted their fragility and reported that No. 2 lacked the power to run with the buckets full. The *Gordon*, for example, worked only 33.6 per cent of the time in May 1900, because a new chain and bucket had to be installed. In June, the dredge was shut down for five days because of continuing bucket-chain problems and recorded 44.8 per cent downtime; the same amount of time was lost during the next month.[21] For September of the same year, the *Gordon* lost 213 of 690 possible hours of work (about 30.97 per cent), of which 49.9 per cent was due to difficulties with the buckets and line. In October, running with 42.9 per cent downtime (more than a third of it because of the bucket ladder), the *Gordon* dug 13,534 cubic yards of gravel, with a yield of 20.06 cents and a cost of 73.72 cents a yard, of which 23.26 cents a yard went for repairs. November wasn't much of an improvement. Working 56.3 per cent of the time, with 46 per cent of stoppage due to buckets, chain, and links, the *Gordon* averaged 2.4 cubic feet per bucket, about half its capacity.[22]

Trouble with the tumblers, the huge pulley-like apparatus over which the bucket chain moved, was also a major cause of lost time. A broken upper tumbler shaft on the *Gordon* in January 1906 cost a total of forty hours. Worse yet, a shattered flange on the lower tumbler of No. 3 electric early in 1907 shut the dredge down for a month or more.[23]

With machinists and an excellent machine shop, Conrey personnel could usually remedy the immediate breakdown. Sometimes, extra men had to be hired for "extensive repairs" to get the dredge "shipshape" again. Occasionally, with the electrics, a malfunctioning motor might require a manufacturer's representative from Salt Lake City.[24]

Some delay problems were the result of tragic human accidents. Records show a stoppage of eight hours for both steam dredges on September 30, 1906, so that workers could attend the funeral of an employee killed while on duty, something that happened at least seven times in Conrey history.[25] Occasionally an effort to be too diligent might bring trouble. In 1909, the pilot in charge of No. 1 electric, eager to recover all possible precious metal, scraped much of a hard granite outcropping most of the way for more than a mile, in the process wearing the points off the lower tumbler flange.[26]

It was not uncommon for a dredge to rip a hole in its hull and sink. When the *Gordon* went down in its pond late in 1906, it took fifty-one hours and forty minutes to get it operating again; and trouble with the moving winch immediately cost a delay of an additional twenty-four hours on the refloated dredge.[27] No. 2 electric dredge sank to the bottom of its pond early in 1911, with ice primarily to blame, but it was readily refloated. At one point in May of the same year, none of Conrey's four machines were operating. No. 1 and No. 2 were undergoing repair, No. 3 was out with a broken ladder, and the big No. 4 was due to be refloated in a few days.[28] Early in 1912, No. 2 went to the bottom a second time and was badly damaged, again mainly because of heavy icing. While No. 2 was out, it was thoroughly overhauled, with the stern strengthened, all at a cost of $22,717.99. Working very hard ground, No. 2 broke spuds on three separate occasions.[29] In the spring of 1914, No. 1 encountered a huge boulder and ripped a large hole in its hull, putting it partly under water pending repairs.[30] No. 1 sank again on the morning of November 18, 1916, when the digging chain caught a well plank and caused a leak in the hull. But the pond was shallow, there was no danger to the machinery or electrical equipment, and the dredge was quickly refloated and resumed its work in a few days.[31]

No. 4 was out for repairs for several weeks in the fall of 1914. For the week ending December 18, 1915, No. 1 lost twenty-three hours and thirty minutes, or about 14 per cent; No. 2 lost forty-four hours and fifty minutes, or 27 per cent; No. 3 was idle all week; and No. 4 lost twenty-four hours and fifty-five minutes, or about 15 per cent—all for a variety of causes, including a broken spud, re-lipping buckets, and digging motor failure.[32]

Some problems could be traced to the manufacturers. No. 4 had trouble with defective plates. In 1917, it had a dozen center upper tumbler wearing plates on hand, which normally would have lasted six months. Because these plates seated the buckets and took repeated shocks and abrasive wear, they had to be made of exceptionally hard steel. This batch, however, proved to be so brittle that some plates broke into small pieces within twenty-four hours. These had not been properly heat-treated, and the company was hard put to get replacements because of a wartime shortage of manganese steel. It fully expected the supplier, the American Manganese Steel Company, to make good on its flawed product.[33]

Running Time and Delays, Conrey Placer Mining Company, Dredges Nos. 1, 2, 3, and 4, January 1 to July 31, 1915

	Time, per cent			
Cause of Delay	Dredge No. 1	Dredge No. 2	Dredge No. 3	Dredge No. 4
Clean-up	1.56	1.32	0.87	1.60
Moving ahead	3.89	6.33	2.75	3.80
Lines and spuds	3.42	3.01	2.25	2.95
Winches	.40	.41	.61	.13
Flume	– –	– –	.48	– –
Ladder	.02	.14	.23	1.07
Tumblers	1.02	1.58	.70	5.91
Hopper	.16	1.14	– –	.87
Grizzly	3.99	3.25	3.03	3.41
Bucket chain and drive	.62	3.05	2.52	2.27
Pumps and piping	.67	2.05	.38	.49
Electrical appliances	1.36	3.13	1.13	1.09
Heating system	– –	– –	– –	.03
Debris	3.66	2.68	.36	.11
Power off	.68	.66	.53	.57
Stacker	.85	1.48	– –	.41
Tables and sluices	.01	.39	.07	.97
Well	.02	.08	.18	.85
Gangplank	– –	.06	– –	.02
Unloading coal	.09	.15	.02	– –
July 4	.71	.78	.78	.71
Hull and housing	.07	1.00	– –	.01
Low water	.03	.38	– –	– –
Winter conditions	– –	– –	42.46	– –
Funeral, President J. J. Myers	.01	.01	.01	.01
Miscellaneous	.02	.01	.02	– –
Total delays	23.26	33.23	59.38	27.28
Total running time	76.74	66.77	40.62	72.72
Total time	100.00	100.00	100.00	100.00

Source: Jennings, *History of Dredging*, p. 28.

Another common problem for Conrey occurred when the dredges worked ground that had previously been mined by hand, sometimes twice—once by white miners and once again by

Chinese. Along with the gold, the dredges picked up pieces of lead and accumulations of black sand and iron. George Edmunds reported in 1917 that the company had on hand 650-700 pounds of lead, mainly bullets, to which amalgam had stuck. He had sent samples to the U.S. Assay Office in Denver, which had put him in touch with the American Smelting and Refining Company in Omaha as a possible buyer. Such a consignment probably would not have brought a profit of more than $350.[34]

Worked-over ground was also honeycombed with old drifts, shafts, and other workings and posed special hazards. Unless extra care was taken, water in the pond might drain away through underground workings, leaving the dredge stranded. One technique used was to work the problem area in three cuts, rather than one, so that the pond would be smaller and more easily controlled.[35] At the same time, the dredge (electric No. 1 in 1917 and 1918, for example) might have to stop frequently to remove old mine timbers from buckets or pond or waterlogged fragments of the timbers that the pumps had drawn into the screen. When No. 1 was working "drifted ground" in the summer of 1918, with disappointing results, Conrey's business manager observed:

> To watch No. 1 operate a person would think we were in a young forest. The ground has been so thoroughly drifted that the old drift timbers rise to the surface constantly, and to such as extent that we have had to keep a man almost constantly pulling them out of the pond and taking them off the deck.[36]

No doubt some difficulties might have been avoided, as in 1917 when there was a minor scare over a fire in coal stored on the then-decommissioned No. 3. Even though all the dredges were electric, the company still bought a substantial amount of coal, which was used for a number of purposes. It was used to heat both dredge personnel and equipment, including the sluices, and was burned in retorting, in the machine shop boiler, in the steam traction engine, and in the prospecting drill. In early June 1917, George Edmunds noted that the wartime situation had prompted the Northern Pacific Railroad people to urge Conrey to arrange for coal early. He anticipated needing 1,000 tons. Edmunds also reported that because of the expense of additional coal bins, Conrey would remove the motors, pumps, and machinery from dredge No. 3 and use its lower deck for storage. In two weeks the first two cars of Montana high-sulphur coal arrived, with more yet to come from the Bear Creek mines near Red Lodge. Jennings, Conrey's consulting engineer, applauded Edmunds's foresight but cautioned him to inform Secretary Stanley of the new plan so that he could adjust the insurance upward.[37]

Three days after Jennings's warning, the forty tons of coal on No. 3 caught fire from spontaneous combustion. Fortunately, crews were able to move all of it quickly, dumping it into the pond before it did any damage to the dredge. "To a man," George Edmunds reported to the Boston office, Conrey's employees "showed the true spirit of interest"; in extremely hot weather and hampered by coal gas and heavy smoke, they had met the emergency head on.[38] The partially burned coal was not worth recovering, and Edmunds estimated the loss at close to $300.[39] Thereafter, Edmunds vowed (and Jennings underlined it from afar) that only a minimum amount of coal—100 tons, it turned out—would be stored on the dredge, in piles only four or five feet high and with adequate ventilation. The rest would be stored under a temporary cover on the tailings pile and distributed to the winter positions of the operating dredges.[40]

Other distractions were indirect and perhaps unavoidable. In 1910, when No. 1 electric was working up the gulch, water from its pond penetrated some old drifts under the county road. In several places, the roadway caved in, leaving it dangerous to the public. The Conrey Company was soon grading a new section of road about six miles from Virginia City.[41] More than eight years later, in November 1918, No. 4 cut the county road where it went past the directors' house, forcing the company to build a temporary road and later replace the old one.[42] Earlier in the same year, No. 4 had been working the tailings left by the *Maggie Gibson*, which had been abandoned and partially dismantled in one corner of the tailings field. The superstructure was still there, however, and had to be cut down, taken apart, and hauled away to the Conrey scrap pile to clear the way for No. 4's swing lines.[43]

A few months later, No. 1 electric would benefit from parts of the *Maggie Gibson*. On July 1, the cable holding the stern gantry snapped on No. 1, and the weight of the stacker broke the four gantry timbers and dumped the stacker onto the tailing pile behind. It took a month, but the gantry was spliced and rebuilt using timbers from the old *Maggie Gibson*. But the long shutdown plus a forty-hour closeup of No. 4 for the Fourth of July contributed to the Conrey financial estimates being "shot to pieces."[44] A forty-hour break over the Christmas holiday was also customary.[45]

When a temporary stoppage was combined with low-yielding gravel or hard-working circumstances, Conrey often felt an immediate economic impact. Late in November and early in December 1917, the company re-riveted the two lower sections of the screen on No. 2. When that was completed, workers did the same on No. 1, along with other repairs preparatory to winter operations. But this meant low yardage, especially because both the electric dredges were running in poor ground at the time.[46] No. 1 had difficult going, working in a high sandbank, forty-seven feet from bedrock, with a great deal of material to move but not much gold. No. 2 was in hard digging thirty-five feet down, with a cemented streak to cut through. In January 1918, No. 2's crew had to take three days to re-rivet the "upper tire."[47] Results from both dredges were disappointing, leaving managers on the scene convinced of the "folly to prophesy" expected dredged returns where No. 1 was concerned. As for No. 2, they reported a bit later, "it seems as if its name should be changed to 'Jonah.'" The dredge was simply not doing well in the tough ground.[48] No. 2 ran at a loss for several weeks that winter and was continued because company officers believed they could recoup the losses on better ground and leave the dredge in a better position for dismantling. Unfortunately, the new ground also proved poor. No. 2's bucket chain and lower tumbler were hauled to No. 1, and the new upper tumbler was destined to go to the shop. No. 1 would be closed down and worked on quickly, then No. 4. But with only two operating dredges, downtime was much more devastating on total results than it had been during the boom years.[49]

Electric power no doubt caused fewer operating problems than steam, although the electric dredges obviously had their share of difficulties. Hennen Jennings calculated that from their beginnings through July 31, 1914, the four electrics averaged 33 per cent lost time, but he gave no analysis of cause of stoppages.[50] As electrician Truman Prier reported in 1909, the distance factor could not be ignored. Conrey bought its power from the Madison River Power Company (later Montana Power) paying four-fifths of a cent per kilowatt in 1913, a higher price than most dredges paid in California. The current came nearly fifty miles at a voltage of 46,000, which was then stepped down at the Conrey substation at Ruby to the 2,200 volts required for the dredge lines. Conrey had built its own transformer station and distribution center at a cost of some $40,000. Prier reported that as dredge No. 1 worked up Alder Gulch, moving farther away from the substation, it lost voltage, making the operation of its motors more dangerous, more costly, and ultimately impossible. The solution, as he saw it, was either to transform the power to 6,600 volts at the substation for transmission to the dredge, where it would be reduced to 2,200 volts, or to install a booster and regulator at the substation to maintain 2,200 volts on the transmission line for an ultimate distance of six miles.[51] To meet the demands of No. 1, which was working farthest away from the substation, the company bought three step-up transformers that transmitted a 6,600-volt current from the station to the dredge, where a step-down transformer brought it back down to 2,200 volts. Because of the transformer aboard the hull, the dredge insurers sought to raise the rates. Edmunds was not sure that they had succeeded, although he later wondered if moving the transformer off No. 1 to the bank would bring better rates.[52]

The electrical system caused other delays as well, the most traumatic of which was the short circuit accident that electrocuted two men on No. 3 in November of 1914.[53] Lightning struck No. 3 during the summer of 1915, but the only damage was a burned-out motor.[54] Rerouting the power cable periodically took some time. In 1917, for example, when No. 4 was passing through tailings to get to better ground, it moved more quickly than usual and required more work to keep the cable in position. At the same time, when No. 1 was working through the old German Bar pit, it had considerable water to contend with and its cable had to be handled with extreme care, taking longer to change than when the dredge was on solid ground.[55] Early in 1918, the big Westinghouse digging motor on No. 4 malfunctioned for the second time in four years. It was replaced temporarily with the smaller spare General Electric motor while a Westinghouse engineer from Salt Lake City came to examine it. Conrey officials concluded that it would be better to keep operating the dredge, even on a reduced basis. If No. 4 shut down, most of the crew would be laid off and would likely find other jobs. The motor problem was apparently with the rotor, but it was not resolved for several months.[56]

If the success of a dredging operation depended on efficiency and as little delay time as possible, properly done gold dredging could be much more predictable than quartz mining, where an estimated 75 per cent of the operations failed. Still, forecasting production required specific ground rules. One reason why a number of early dredge enterprises collapsed was their failure to test the ground adequately before installing expensive machinery. To determine accurately the value and character of the gravel and the bedrock, either shafts or drill holes must be sunk and the results considered along with such factors as possible fuel or power supply, labor conditions, transportation, water availability, physical features, cost of land, and climate. Shafts were generally too expensive, so the value of dredge ground almost everywhere was computed from sampling with drills, generally steam-driven. Constant care was necessary to provide accurate sampling and to prevent the salting of samples, which occurred in dredging just as it did in all other types of mining.[57]

From the beginning, the Conrey people followed a careful policy of advance testing. As early as 1896, J. C. Sloss supervised preliminary exploration by dropping test shafts and drilling with a Keystone rig, the Pennsylvania-made churn drill that became standard in the industry. In 1897, the company spent $5,251.18 for a "Prospecting Drill," probably a Keystone. Three years later, the company was test drilling at the mouth of Alder Gulch, looking for a second bedrock below the first, and nearby ranchmen were soon also prospecting their land with drills.[58] Late in 1903, when Conrey directors authorized manager Baier to order lumber for a new dredge, they also empowered him to either purchase a new traction-type drill or provide a new boiler and engine for the old one.[59]

On Conrey ground, Keystone prospecting involved dropping test holes at regular intervals that were farther apart than became customary in California. On one part of the Conrey property, holes were staggered 150 feet apart on lines that were each 200 feet distant from another. On another part, rectangles were measured 330 by 660 feet and holes were drilled in the corners. At least during the steam era, dredge recovery showed about a 5 per cent loss in the tailings over the prospecting analysis.[60]

The company also kept one man busy part of the time surveying the ground, so that each month's yardage could be ascertained and entered on the property maps. Normally, the engineer at Ruby did the job; but if the staff was short-handed, the county surveyor might be hired for several days at the beginning of the month or Leslie Gilman, the clean-up man, might be called in to handle the work.[61]

In an effort to estimate both yield and longevity of operations, surveying and prospect drilling were ongoing pursuits throughout the life of the company, and a drill crew was maintained for that purpose and to test potential new property. Even as late as July 1917, the Boston office was receiving estimates of ground yet to be dug by No. 2 and No. 4 based on drill holes sunk. Ground left for No. 4 was believed to be 1,025,000 cubic yards at a depth of 26.3 feet with a projected yield of 15.5 cents a yard. No. 2 had 1,160,000 cubic yards remaining at a depth of 29.9 feet; the return was estimated at 13 cents a yard.[62]

Conrey officials were always on the alert for new ground, which sometimes led to drill prospecting in contingent or even more remote areas. The company added new Alder Gulch property with some regularity until about the time of World War I, sometimes acquiring ground outright and sometimes taking an option until prospecting could be done.[63] When the Conrey people bonded property owned by the Alder Gulch Consolidated Mining Company in 1907, it first sent W. W. Payne and his crew to survey it and then brought in Newton Pilger and a Keystone drill to test it systematically.[64]

The company was also interested in running tests elsewhere. In the winter of 1908-1909, Kammerer took a drill crew into Elk Park near the Great Northern station east of Butte, but cold weather drove them out until the next fall.[65] The company also had an option on 600 acres in German Gulch, about sixteen miles west of Butte; and in late 1909, Conrey sent in a Keystone drill to test it. During the following spring, Charles Kammerer dropped more test holes on the Conrey property in Alder Gulch, and the company spent at least fifteen hundred dollars to send San Francisco engineer Ross Hoffman to do additional drilling.[66] In 1910, Kammerer sampled placers at Pole Creek owned by A. W. Tanner, and four years later Conrey sent John Batten to Oregon to test a large tract.[67] In the last summer and fall of 1915, Conrey sent Batten and a drill crew to Barton Gulch for several weeks of testing, then to the area near Townsend.[68] None of these possibilities proved to be promising enough to follow up on. In 1917, Edmunds talked about prospecting of the Gravelly Range, which Nathaniel Shaler and attorney William Clark had explored in an earlier era, but nothing came of it. And in 1918, with its ground about used up, the company decided to not concern itself with offers of new sites near Missoula and in Canada.[69] But in general, throughout most of its active life, Conrey kept its drill crews busy prospecting ahead of the dredges or testing potential new fields.

Conrey's experience bore out a truism of twentieth century dredge mining: namely, that the greater efficiency and larger capacity of modern dredges steadily decreased working costs per cubic yard, despite shorter working hours and higher pay scales. On some of the early boats, labor alone cost as much per cubic yard as total operating charges at the time of World War I. Between September 1, 1906, and July 31, 1915, the Conrey electric dredges dug more than 31 million yards, with an operating cost of 5.91 cents a cubic yard plus indirect expenses (such as insurance, taxes, Boston expenses, legal fees, and roads) of 1.05 cents, or a total of 6.96 cents per cubic yard. Working costs began with 9.79 cents in 1906 and were reduced to 5.36 cents a cubic yard by 1914, a year in which the company dug more ground than it had during the first four years combined. Figures show that 1.76 cents a yard, or nearly 30 per cent of the cost, had been due to material and labor for repairs. Power averaged 1.32 cents, labor 1.37 cents, and chain upkeep 1.25 cents.[70] Because it was more modern and more efficient, although by no means the most up-to-date dredge in the fleet, in May 1915 the Conrey No. 3 was able to dig through some of the tailings from the old steam dredges and recover from 2.05 to 3.73 cents a cubic yard.[71]

More advanced than their steam predecessors, the electrics led the way in "navigating Alder Gulch." But none of the Conrey dredges found clear sailing. They were ponderous, yet fragile machines, whose log books charted the ups and downs of their daily voyages, tempered by numerous obstacles and problems.

Conrey land as it appeared before dredging in 1900, with the company's first two steam dredges visible in the center background

The *Maggie Gibson* (right), the inspiration for Conrey's fleet of six dredges, and Steam No. 1 (left) in about 1900

Steam dredge No. 1 under construction in 1899, with front gantry for hoisting the bucket ladder nearing completion at center

Hayward "orange peel" machines dumping gravel into a long sluice in the late 1890s

Dredge No. 1 in operation in August 1901

Steam dredge No. 1, which was built in 1899 and dismantled in 1907, showing sluice at rear

The Conrey Company's machine shop in 1905

The Keystone prospecting drill, with John Batten (center) panning, Winchester Dickerson (right) rocking, and Nelson A. Brissette standing at the drill

Conrey's Keystone prospecting drill, used to test the gold value of Alder Gulch property before purchasing

Electric dredge No. 3 changing from "1904" bucket-chain to a "1905" chain in 1908

Steam dredge No. 2 under construction in 1901

Setting the machinery on steam dredge No. 2 in 1902

Steam dredge No. 2 under construction near Ruby in 1901

Steam dredge No. 2 launched but not yet digging in 1902

Electric dredge No. 1 under construction in 1907

Construction begins on electric dredge No. 3

Working on the hull of electric dredge No. 3

The dredgemaster in No. 4's pilot house in 1912

WINCHESTER
W FEBRUARY 1912
1 2 3
4 5 6 7 8 9 10
11 12 13 14 15 16 17
18 19 20 21 22 23 24
25 26 27 28 29
GUNS AND CARTRIDGES

Conrey worker retorting in the open air in 1902, with Ruby dance hall in the background

One of Ruby's two saloons in 1904

Fourth of July celebrations at the company house in 1904

The mooring winch and pumps of electric dredge No. 3

Dredge No. 3 near Ruby in 1908, shown with its bucket chain out of the water

Veteran clean-up and retort man Leslie Gilman (left) in 1912 with gold bricks worth about $5,000 each

Machinist William Bryant with upper tumbler casting from electric No. 2 on the planer at the Conrey shop

Electric dredge No. 4

Chapter Seven

Of Men and Machines

The Boston office set the Conrey company's general policies, but it rarely meddled in day-to-day operations. Directors may have been conservative, but many of them knew first-hand what mines and mineral investments were all about. In the early days, the company's direction was in the hands of able men like Gordon McKay, Nathaniel Shaler, and Philip Moore—men who had considerable knowledge and expertise in business, technology, and mining. There is no evidence that McKay ever visited Ruby, but Moore spent a good deal of time there and Shaler made the long trip at least three times. For Shaler, getting away from the academic environment was a therapeutic experience. Whatever he was in Cambridge, he would say with a smile, he was a great man in the Rockies. Genial and outgoing, the trim, sandy-haired professor was delighted with many of his more rustic western associates and with the opportunity to give eastern friends or students a guided tour of the "dredge boats" in action.[1]

After he replaced Moore, both as consulting engineer and on the board of directors, Hennen Jennings became a major factor in corporate decisions for Conrey's Montana operations. Although not in residence, on a number of occasions Jennings either sent skilled California engineers to inspect operations or made the journey himself. Conrey's directors and their technical consultants were remarkably temperate and level-headed, always amenable to suggestions from the field and giving considerable discretion to those on the scene at Alder

Gulch. Fortunately, the gulch still held a profitable supply of gold, and the men in charge at Ruby were competent and conscientious, although by no means exceptional.

From at least early 1901 until his death in 1921, Frank F. Stanley served as the company's secretary and treasurer, coordinating paperwork and performing all the central office functions. From his office overlooking Boston's Inner Harbor, Stanley dispersed payments for major equipment, handled the distribution of dividends, and wrote the annual reports. It was Stanley who used preliminary profit-and-loss reports from the manager at Ruby to help "in the compilation of numerous tax returns," which for 1917 broke down to $10,318 for state and federal taxes and $9,127.81 for county taxes. But Stanley sometimes lacked adequate information. When Edmunds sent an elaborate valuation of all thirty-five company buildings, ranging from buggy sheds to the big machine shop, Stanley was unable to determine the Montana rate for depreciation and Edmunds could provide only a partial answer.[2] Stanley also took care of all insurance on the Montana property, including dredges, and retained the policies in Boston. Premiums were not inconsequential: for 1918 they totaled $5,454.20, with insurance on the big No. 4 alone running $2,494.82.[3]

Management in Alder Gulch was usually in the hands of men with extensive dredge or business experience, most of whom had come up through the ranks at Ruby. Following Philip Moore, Julius Baier directed Conrey operations in Montana until his death in May 1905. Having both Boston and St. Louis connections, Baier was apparently chosen because of earlier associations with McKay, or perhaps Moore. Young and vigorous, Baier had been seasoned in the Colorado placer fields, although how much dredge practice he brought with him is uncertain. Baier arrived in Alder Gulch in October 1900 to take charge of planning steam No. 2 and the new electric No. 3 and the operation of the *Gordon*. When a young Harvard graduate met him there in the autumn of 1902, Baier was "all over the boat, directing the men," but he still found time to explain everything to the visitor. But Baier soon fell ill, and thermal treatment at Hunter's Hot Springs near Livingston, Montana, provided no cure. He died at the age of forty before the first electric dredge was completed.[4]

Charles Kammerer was named superintendent following Baier's death, and Adolph H. Wiseman became acting manager and then manager—both at salaries of $2,500 a year. W. F. Ward became the company engineer at $180 a month, and M. H. Gerry, Jr., was named supervising engineer to oversee the completion of No. 3.[5] Wiseman, who seems to have had some link with Nathaniel Shaler, had given up the study of law in St. Louis in 1901 and came to Ruby as bookkeeper and cashier of the Conrey Company.[6] He was in charge of all operations, including dredging, machine shop, and bookkeeping. Kammerer, who had been "foreman at the dredges" when steam No. 2. was under construction, specifically directed the dredging work. Ward, a graduate of the Colorado School of Mines in 1902, did surveying and design work on the new No. 3 in 1903-1906 and was considered "energetic, capable, honest but somewhat conceited and opinionated."[7]

Ward soon left Alder Gulch, and Truman D. Prier took over his work in addition to generally supervising the clean-up and some electrical tasks.[8] Wiseman remained as manager until August 1908, when he resigned to resume his law studies in Seattle; the company bought his piano, gramophone, and books as a nucleus for an employees' reading room.[9] Admitting that he "did not do as well as I should have liked" at Ruby, partly because of lack of technical training and "some degree of temperamental unfitness," Wiseman borrowed money from Hennen Jennings and eventually became an attorney on the Coast.[10] Wiseman's successor was Charles Kammerer, who was promoted after working three years as superintendent.[11]

Kammerer spent nine years as Conrey manager at Ruby before being enticed into war industry work in California in 1917. He had an excellent reputation as manager, although Jennings confided to Charles Francis Adams that Kammerer's leaving was not a great disaster. Near the end of his stay, Jennings reported, Kammerer had a record of absenteeism and set a bad example to employees.[12] Truman Prier, a thirteen-year veteran with the company, was appointed acting manager until Jennings could get to Ruby and appraise the situation.[13]

Once on the scene, Jennings was convinced that Prier could not be put in charge. A family man, Sunday School superintendent, and long-time company employee as draftsman, electrician, and engineer, Prier apparently succumbed too often to the blandishments of John Barleycorn. Mining engineer Thomas Leggett disagreed. Leggett visited the Conrey ground for eight days in 1917 and concluded that Prier was an able man, repentant about his drinking, and deserving of another chance.[14]

But Jennings had his way. Under his reorganization, a kind of local triumvirate emerged, with Prier in a limited and temporary role. Forty-two-year-old Oscar O. Sweeny, a dredgemaster of long experience, became dredge superintendent at a salary of $275 a month; he also had the use of a "Ford runabout" for company business. Sweeny was put in charge of dredge and machine shop operations; he also approved all warehouse requisitions and did all hiring and firing of employees, but only after consultation with the business manager. The business manager was to be George H. Edmunds, a Massachusetts native who had handled business affairs for the company for more than half a dozen years. Part of his charge was to draft a monthly report from information supplied by the dredge superintendent and the engineer, which would be initialed by all three before it was sent to the Boston office. Edmunds would sign all checks, with countersigning by the engineer or by Superintendent Sweeny. Prier was made the company engineer at a salary of $225 a month, plus a house and other perquisites, but his was not a permanent position. Because Conrey planned on winding down its affairs in a few years, according to Jennings, there was little engineering activity "and thus no justification for the company retaining high class talent in this department beyond the end of this year." Prier was placed in charge of electrical matters and was to finish drawings and estimates for the proposed renovation of No. 1 electric dredge. He would also countersign Edmunds's checks and either he, Edmunds, or Sweeny were to be present at clean-up and at retorting time to verify the bullion return. Should there be disagreement among the three, final arbitration was to be made by the company's Virginia City lawyer, William A. Clark.[15]

Jennings also imposed some "conditions" on the new management. He told Prier that if he "cheerfully and harmoniously" handled his duties and met "other conditions to be later mentioned," in light of his excellent earlier contributions to the firm Prier would receive a "substantial bonus at his departure." These "other conditions" included the pledge of both Prier and Edmunds to refrain from using alcohol at Ruby and Alder. If either was shown to have broken this promise, Jennings warned, "their immediate resignation is in order and will be accepted." Prier went even further that he had been asked, pledging complete abstinence while on the Conrey payroll. Even dredge superintendent Sweeny, a "moderate drinker," agreed to help curb excessive drinking among the crews, but he argued that he "would lessen his influence among his men if he now should give up entering a saloon." Department heads Prier, Sweeny, and Edmunds should present "upright clean and wholesome public example in private life" to all employees, although Jennings hoped that the impending state prohibition of alcohol at the end of 1918 would also help. The company had refurbished a hall for movies and dances and had fitted out an old log cabin for "a club resort," but it was up to the heads to set the standards.[16]

Truman Prier was not part of it. After the managerial rules were laid down, Edmunds reported, Prier continued "fighting a great battle with himself," blamed others for his problems, lost interest in his work, and worried needlessly about his finances.[17] Prier was amenable to breaking in Ambrose H. Burroughs, Jr., a young Columbia Mines graduate who would have limited engineering duties. But Prier was slow in finding another position. Because of his chronic absenteeism and an inability to concentrate on his job, the company soon terminated his employment. Prier immediately invested in a combined movie-dance hall-confectionery in Whitehall and took with him as assistant the pilotman on dredge No. 4.[18]

Jennings hired Burroughs even before Prier left, to the chagrin of Business Manager Edmunds, who doubted that there was enough skilled work to keep him busy. Once on the scene, Burroughs soon made it clear that he was unhappy with his situation. His work, he said, was not that of a trained engineer. He had no private office. He did not care for the physical aspects of cleanup, which he insisted was done too frequently. He laid down his own rules for planning prospecting and the gulch survey and informed officials in Ruby that he would leave when he found another position. Edmunds was "annoyed at his spirit and lack of interest and co-operation in recent months" and thought that unless there was "a radical change" in the progress of Burroughs's work and his attitude, "it is a waste of money to await his pleasure about leaving." Eventually, Edmunds reached a more amicable relationship with Burroughs, but it was late in coming. After a year Burroughs left to accept a commission in the U.S. Army Engineers.[19]

Edmunds and Sweeny directed on-the-scene operations until the company was dissolved in 1922, with the millionaire Hennen Jennings as a roving consulting engineer until his death in 1920, at which time the Ruby offices flew a borrowed flag at half-mast and closed down the dredges for an hour during his funeral. When Conrey was no more, Edmunds remained in mining promotion and as late as 1936 spent a season managing a dredge in Alaska.[20]

Like other types of mining, dredge mining was a specialized business and employed both skilled workers and ordinary laborers. In general, the steam dredges required more manpower than the electrics. In 1903, according to the local press, the *Gordon* employed twelve men to cover day and night shifts (two more than on the *Maggie Gibson*), headed by foreman Charles Kammerer.[21] Other sources indicate that the old steam dredges required eleven men each in twenty-four hours.[22] In July 1906, sixty-five men made up the Conrey payroll, which was temporarily larger than normal because of construction on No. 3. At that time, a regular steam dredge crew, day shift, consisted of pilot, engineer, fireman, flume man, deck hand, coal passer, and lineman. The night shift was smaller and included only the pilot, engineer, fireman, and flumeman. These eleven men ran the boat on two almost around-the-clock shifts, with thirty-minute stops at noon and midnight for oiling and other regular maintenance. The dredges ran a twenty-three-hour day, barring breakdown time.[23]

The new electric dredges required slightly less labor than their steam predecessors. By late 1911, with all four electrics operating, the Conrey Company had eighty-five men on its payroll, and several hundred people in Ruby and the surrounding area depended on the company for a living.[24] By this time, a crew included a dredgemaster (day shift only), a pilot (or pilotman), two oilers, a lineman (day shift only), a teamster (day shift only), and two laborers (day shift only).[25] According to state reports in 1908, pilots received $5.00 a day, oilers and flumemen received $3.50, and laborers received $3.00 for nine hours.[26] Five years later Jennings indicated that eight-man crews ran dredges No. 1 and No. 2, while ten men were employed on each of No. 3 and ten on No. 4. In addition, he reported, four more were at work in the office and dwellings, two worked at the transformer station and on the pole line, three were in the clean-up crew, twenty-two fell into in the category of laborer, and another ten fell under the miscellaneous heading.[27]

The dredgemaster was akin to the shift boss of an underground mine. It was his job to keep the machine or machines in operation and to keep personnel moving, and he had to make sure that pilotmen kept production up and missed no bedrock. He had to be adept at handling a dredge and at the same time be a master mechanic, so that in the case of an accident he could improvise and get the machinery turning with as little lost time as possible. The pilotman (often called the winchman) ordinarily ran the boat. A good pilotman could understand almost by feel how to raise one spud and drop the other at the end of a swing without stopping the digging process. With an ammeter to warn him if the line encountered an obstacle or if the buckets ran out of ground, he could presage the event. The lineman usually worked day shift only, but his work required considerable skill and exposure to disagreeable weather conditions. "A poor lineman can cause the shutting down of a dredge frequently," Edmunds and Sweeny reported in 1918.[28] The oilers kept the machinery in shape. To keep the deck clean, deckhands broke bedrock off when it stuck to the bucket lips and kept a constant eye on both stacker and screen for any evidence of clogging. Shoremen dug "deadmen"—heavy logs that served as anchors for the lines from the dredge, handled supplies, and did an assortment of chores.

Particularly important was the special clean-up crew—the men who periodically picked up the gold from the tables in the dredge, processed it, and deposited it. Clean-up was understandably a duty couched in caution and security measures. Here was the fruit of the hunt for treasure. Here the gold, picked up with quicksilver to form an amalgam was tangible and transportable. Access was limited and the tables themselves were behind lock and key, available only to the clean-up crews and sometimes to the dredgemaster. The dredgemaster on No. 4 had a key in

1917, but the dredgemasters on No. 1 and No. 2 did not, a situation George Edmunds believed ought to be altered so that all dredgemasters had access. After clean-up, he argued, repairs often had to be made on the tables, sluices, and distributing trough under the screen. To get the valuable amalgam off the dredge as soon as possible, the clean-up men brought it promptly to the office or the retort shop, leaving the dredgemaster and his crew to make the needed repairs and then to lock up. There were also times when the spray pipes playing on the tables clogged, and the dredgemaster needed a key to see that they were cleaned.[29]

On the Conrey dredges, the clean-up crew was made up of veterans who had performed the chore for years. Leslie Gilman and Nelson A. Brissette had both joined the company in 1903, and both were still handling clean-up chores in 1917. Gilman would continue until the concern was disbanded.[30] The son of pioneer homesteaders, Gilman had been born in a log cabin near Alder and had caught the eye of Conrey managers after he had invented a horse-powered merry-go-round. Through the years, he worked as an electrician and surveyor and he had cleaned the gold amalgam out of the dredges once a week or so, purified the amalgam in the two Allis Chalmers retort furnaces in the retort house, and transported the forty-five-pound gold bricks he cast to the express office in Alder. He carried as much as $40,000 a week during the lush days when all four dredges were on good ground. "I made enough gold bricks during those years to build a house," Gilman commented.[31]

In 1918, one relatively new engineer, disgruntled to some extent that he had to do tasks that he felt were below him, complained when he had to help with the clean-up. He charged that the job was done too frequently because management liked to show its authority and keep the clean-up crew busy. Business Manager Edmunds pointed out that in the boom days, with the full quartet of dredges operating, engineers and even company managers helped with the clean-up in all types of weather.[32]

Special precautions had to be taken at night also. In order to gauge the water level, the oiler was required to take soundings and measure the bank above the pond from shore. These soundings had to be made when the dredge had finished its cut to bedrock and was preparing to move ahead, so the job might need to be done at any time, day or night. If the sounding was taken at night, according to the strict rule of dredge operations, the gangplank had to be elevated higher than it was during the day to prevent anyone from sneaking onto the dredge. No doubt this was at first a safeguard to the amalgam, although during World War I conservative dredge men also saw this as a precaution against possible depredation by the radical IWW element in Butte, for example.[33]

Perhaps tight security was not such a bad idea, as an episode of 1908-1909 indicated. When someone noticed that Arthur Legree, a $3.50-a-day oiler on No. 3, spent more than a day's wages each night in Ruby saloons, Manager Charles Kammerer suspected him of robbing the flume. After keeping him under surveillance for several nights, at four o'clock one morning Kammerer saw Legree shut down the pumps and "delve into the flume with his hands." A noise alarmed the oiler, and he proceeded into the hull. A subsequent search of Legree's house turned up two steel retort pans, and he was arrested, protesting his innocence. Free on $2,000 bond, Legree jumped bail and disappeared; his wife filed for divorce, naming a young Ruby woman as correspondent. Some time later, he was apprehended in New Jersey working on a farm, and his sister in Butte was arrested for receiving the stolen amalgam. Sheriff Nicholas Tranfler traveled to New Jersey, paid over three hundred dollars in reward money, and started home with his prisoner, who was wearing a twenty-five-pound "Oregon Boot." While the sheriff enjoyed a shave in a Chicago barbershop, Legree slipped the "Boot" and vanished. Six months later, he was apprehended in Albuquerque, employed in a coal mine; he was brought back and sentenced to ten years in the penitentiary at Deer Lodge.[34]

Most of Conrey's labor was hired on a permanent basis, although occasional local artisans were hired for specialized jobs on a short-term basis. For example, Fred Malone, apparently from Virginia City, was hired in the spring of 1904 to do some tinning work.[35] E. M. Smith, also from Virginia City, did some special electrical work for the company in its shop in August 1906. He was retained on a more permanent basis a few months later and eventually left in 1909 to open his own automobile garage in Virginia City.[36] Occasionally, the company contracted with an outside firm for major tasks, as in 1915 when it arranged with the Mountaineer Welding Company to use acetylene torches to cut off the manganese-steel lips of eighty-four buckets that did not have adequate clearance. A total of 140 lineal feet was cut in six hours, a substantial achievement.[37]

Many employees were quite mobile. Some wintered elsewhere, then returned in the spring to jobs on the dredges.[38] Others considered Alder Gulch a temporary stop and moved on to other employment elsewhere in the West, often following new mineral rushes or striking out in business for themselves.[39] Other employees "moonlighted" while holding down jobs on the dredges. Leon Dessault, who would lose his life in an accident on No. 3 electric, was developing a copper ledge after work hours; H. L. Doherty, pilotman on No. 4, leased a ranch at Alder "and will grow parsnips, peas and pumpkins as a side issue this summer," according to the local editor.[40]

Conrey employees knew that mining was a dangerous profession. Gold dredging avoided many of the perils of underground work—blasting, caving of stopes, poisonous gases, and explosions—but it was still a hazardous occupation. Powerful machinery could maim or kill, drowning in the dredge pond was not uncommon, and the application of high voltage electricity added another lethal risk. While corporate managers naturally sought safety in their operations, neither state nor federal law laid down much in the way of specific regulations until the very end of the Conrey period. Whether the Conrey record was better or worse than its their counterparts elsewhere is conjectural. Terrible accidents did occur with all too alarming frequency.

In the summer of 1902, C. C. Foster was working on the flume and dumping apparatus on the No. 2 steam dredge when the machinery started up. Foster, a popular man with a wife and three children, was struck on the head and knocked unconscious; miraculously, he survived, although doctors despaired for a time.[41] Only three months later, the *Alder Gulch Times* carried the lurid headline "William Schneider Instantly Decapitated at the Dredges." Forty-five-year-old Schneider, an engineer in the No. 2 pilot house, was dragged into the massive machinery in the lower workings when his clothing was caught. A coroner's jury ruled that his death was accidental and caused by his own carelessness.[42] The company may have been exonerated of blame, but a year later the directors voted $300 to his widow.[43]

J. Sweeny, a pilot on the *Gordon*, was more fortunate when he caught his hand in a pulley belt in the summer of 1902. It was badly lacerated but not broken.[44] In the winter of 1905, John Moenster somehow fell from the coal scow at one of the dredges and was drowned, although there were no witnesses.[45] A few months later, a heavy piece of iron crushed the toes of J. H. Monaghan, who was at work on one of the Conrey machines.[46]

Two major accidents occurred in 1908, one of them fatal. William F. Kammerer, the brother of Conrey superintendent Charles Kammerer, had been an experienced placer miner for twenty years. In his first week as flume man, he was struck on the head by the steel handle on the winding drum of the winch that controlled the flume cable. He died almost immediately, leaving a wife, four children, and a $4,000 life insurance policy.[47] In May, in "another frighful accident" on No. 1, which the Marion Steam Shovel Company had not yet been turned over to Conrey, oiler Charley Cordell slipped on a ladder. While grabbing for something to hang onto, his right arm was severed by two large bevel gears. With "wonderful nerve and rare presence of mind," he walked up two flights of stairs to the pilot house and survived.[48] Conrey directors decided that if Marion would contribute to the injured man, they would too, and in 1910 Cordell received $1,000.[49]

Working on electric No. 2 just before Christmas 1909, Albert Nelson lost an eye after being struck with a piece of flying steel when he and another man were cutting rivets on a bucket.[50] At about the same time, an employee named Paxton was accidently killed. The directors absolved the company of all blame and offered the family $1,000 in compensation. When suit was brought, however, Conrey settled out of court for a larger, unspecified sum.[51] O. O. Sweeny, superintendent on No. 1, suffered a broken ankle and badly torn ligaments when his foot caught in a cable in early 1911. In an accident on No. 2 a few weeks later, Charles Kinkaid lost the end of one finger and Henry Petit lost three fingers.[52] Six months later, while the dredge was shut down for repairs, on No. 1, oiler Harry P. Fitzpatrick, a man of four years experience on the big machines, was caught in a moving belt and killed instantly. "Nearly every bone in his body was broken," the local paper reported.[53]

A year later, James Duncan, Jr., was badly injured on No. 4 when he fell sixteen feet from the grizzly to the deck below and struck the back of his head.[54] And 1914 was a particularly bad year, beginning with a smashed thumb for Joe Dionne and a broken collarbone for Herbert Wiles, whose clothes were caught in the machinery on No. 4.[55] In November, "one of the saddest accidents in the history of the Conrey Mining Company" occurred when Leon Dessault, a long-time pilot on No. 3, and employee Jacob J. Hagenach, were both electrocuted. Dessault, a man of about sixty, had each hand on a lever, controlling the digging and shifting mechanisms, when a short circuit passed 2,200 volts through his body. When Hagenach came to his aid, he accidently touched a lever and was himself electrocuted.[56] Because of the expense, the company carried no accident insurance. Conrey's attorney, William Clark, believed that a settlement of five to six thousand dollars would be appropriate in the Dessault's case, but Charles Kammerer believed that the figure was too extreme.[57] Subsequent investigation revealed that the digging controller had never been grounded; in fact, it had no ground wire attached. There was no question that the company was liable for the accident to Dessault; but because Hagenach's work was on shore "and he really had no business" on the dredge except when he saw Dessault stricken, Kammerer concluded that Conrey was not legally responsible for his death. Kammerer paid Dessault's funeral expense and the cost of shipping Hagenach's remains to Pittsburgh. Despite rumors of a lawsuit on Hagenach's behalf, he believed the company ought to settle justly with the Dessault family first. Kammerer prompted court action and that the case had been settled out of court.[58]

There were plenty of lesser injuries over the years. In the fall of 1915, Thomas Anderson received several broken ribs in an accident in the machine shop. A few months later, Ray Gilman, one of the company's electricians, was repairing wires at the top of a pole when it broke and fell on him, breaking his leg in the process.[59] In 1917, one employee had his finger badly mashed while repairing a motor, and another hurt his spine when he fell and heavy steel tumbled on him. An oiler in a rowboat below one of the dredges received a broken rib when he was struck by a chunk of falling mud.[60] Late the next year, in the midst of the influenza epidemic, with at least nineteen Conrey employees suffering from the malady, an electrician cut the back of his hand badly, severing a tendon. He required medical attention, first at Sheridan, where the doctor had the flu, then finally at Twin Bridges.[61]

What special safety precautions managers of the Conrey Company took are not clear. As of mid-March 1910, at least, employees at Ruby were not covered by accident insurance, although directors had authorized Secretary Stanley to investigate the possibilities. Presumably, the Montana Workman's Compensation Act, which went into effect in 1915, gave Conrey workers at least some protection.[62] Nobody working on the huge dredges amidst large gears and moving belts and cables could ignore the danger, and the added peril of high-voltage electricity was always present once the dredges were electrified. In 1918, Hennen Jennings sent officials at Ruby a copy of the California statutes on safety devices on gold dredges.[63] But dredgemen were like other miners: risk invariably went with the job.

For two decades or so, relations between the officers of the Conrey Placer Mining Company and their employees in Montana were generally excellent. Wages and working conditions were competitive, and the company had a good reputation. In 1901, when the dredges were put into operation, the directors authorized increased wages of fifty cents a day to the levermen.[64] But 1906 brought at least a veiled threat of a strike, which prompted Hennen Jennings, who had just taken a hard look at the Conrey operation, to recommend that the company pay good rates and conform to the established work hours of the area and profession. Dredge workers wanted to receive the Butte scale—an eight-hour workday with pay for twelve, a change that would mean three shifts instead of two and an average additional labor cost of about one thousand dollars a month. Jennings recommended this upgrade for workers on the electric dredges, but he resolved "that the Company submit to no dictation as to the number of men to be employed or what dredges are to be used."[65] By Jennings's standards, daily wages were already high: pilots received $5.00; engineers, $4.00; firemen, $3.50; flumemen, $3.50; deck hands, $3.50; coal passers, $3.50; lineman, $3.00. Hourly wages were forty cents an hour for machinists and blacksmiths, forty-five cents for carpenters, and thirty cents for laborers.[66]

Meeting in Boston in August 1906, the directors were cautious but not unreceptive. They instructed A. H. Wiseman, the manager at Ruby, to delay demands for the eight-hour day until the three electric dredges were operational. He was not to grant a change in hours without the board's sanction, although he was authorized to give labor in the shop a nine-hour workday when No. 3 started up.[67] Soon, however, eight-hour shifts became standard on Conrey dredges, which now ran around the clock with no scheduled stops, except for periodic clean-ups.[68]

Beginning in 1909, Conrey gave its employees a bonus "for faithful and efficient service." The bonuses ranged from five dollars for new workers to more than a month's wages for old-timers, with a total of $5,000 being prorated among those on the payroll as of December 31, 1908. "This is the first time in the history of the country that a mining company has taken such magnanimous action," the *Alder Gulch Times* editor announced.[69] The practice continued intermittently over the years; and when a third bonus was voted at the end of 1910, $12,500 had been distributed, a bonanza not lost on the recipients.[70] Certainly, corporate officials were proud of their relationship with the men in the field. As Secretary Frank Stanley wrote one of the shareholders in 1915, "We have a good class of men at the mine and as they are well paid and housed, they are content, and we have no labor strikes."[71]

The manager in the field did most of the hiring. George Edmunds often wrote consulting engineer Jennings and sometimes company secretary Stanley about appointments, but Jennings maintained a "hands off" attitude in hiring matters, except for those at the engineer level. He insisted to the Ruby managers, as well as to the august Charles Francis Adams at Harvard, that one of the fundamentals of good management "is to leave the selection of subordinates to the Manager, especially in such a small place as Ruby."[72]

World War I soon brought changes. War-related activities on the Pacific Coast drew off workers and raised the level of wages everywhere. Western mining and agriculture found themselves in fierce competition for labor with each other and with countless war-time industries elsewhere, with the rivalry becoming even more intense after the United States entered the conflict. Even in remote Montana (or perhaps because it was in remote Montana), Conrey felt the impact in tangible ways.

The company lost men, most notably Manager Charles Kammerer, who left in 1917 and joined the Union Construction Company, a San Francisco dredge manufacturer under contract to build ships.[73] The monthly report for April 1917 was late getting to the Boston office because a veteran bookkeeper had enlisted to avoid being drafted.[74] The head electrician left, and the firm's best machinist went to Mexico.[75] Edmunds and Sweeny believed that the company was fortunate to have so many older, sedentary employees and argued against closing down dredges and laying men off lest they find other jobs and not return when work commenced again.[76] The two managers especially noted the loss of laborers, but there were also some oilers and two pilotmen, who had been attracted by extraordinary war-time wages and rising copper prices that boomed the Butte mines.[77] One man left to take charge of a ranch at an inflated salary, and high grain and cattle prices attracted a few older employees who bought or leased land and became entrepreneurs.[78] "Companies seem to be bidding against each other," the Ruby management complained, "offering increased wages, shorter working hours and in some cases bonuses if employees remain a certain length of time."[79]

The dredgemaster, the pilotmen, the oilers, and the lineman were all professionals, and most of them had been with the company for years. Because of their age, special skills, and marital status, they were much less mobile than the laborers, who tended to go to wherever wages were immediately highest. Of twenty-seven skilled dredgemen on Conrey payrolls in mid-1917, when only three boats were running, twenty-six were married; only seven of twenty-one laborers were married. "The usual laborer does not care how often he moves, but will go where he can make the most money for the time being regardless of the future," Edmunds and Sweeney reported to Jennings.[80]

The war-time boom also favored generalists like electricians, bookkeepers, and machinists, whose expertise might be applied in many industries or businesses. Machinists in particular were in short supply during the war years. When Conrey managers wrote the Marion Steam Shovel Company for help, Marion officials replied that skilled operators were scarce in Ohio as well and "that the automobile concerns were robbing them of their machinists." In desperation, in the summer of 1917 the company put Axtel Brynie, dredgemaster of No. 1 in charge of the machine shop—at dredgemaster wages. Brynie had worked as mechanic and blacksmith before advancing to pilotman and then dredgemaster, and was willing to take permanent charge of the shop so that his children might go to school in Ruby. Charlie Shott had been with the company since 1900 and was the head blacksmith in 1918. Because of his advancing age and the loss of one eye from a flying piece of steel, Shott could no longer do the physical work he once did; but under his supervision, his helpers got it out.[81] Edmunds and Sweeny hired a machinist from Denver with prior experience at Bethlehem Steel, but in general they found it difficult to attract skilled personnel from Colorado or from Butte, where wage scales were higher than those paid at Ruby. One man contacted in Butte wanted six dollars for eight hours; one in Denver wanted 62.5 cents an hour, plus at least part of his travel to Montana. Neither came, nor did a Californian who was approached. Two other machinists had been promised by employment agencies in Colorado, but Sweeny and Edmunds were not optimistic: "they have not reached their destination, probably obtaining employment enroute."[82] Ruby managers tried to entice electricians away from the Montana Power Company and from Butte, but without luck. Finally, they did attract Fred Strouse from Chico, California, a man who was not only a teetotaler but who had his own full kit of tools.[83]

Even the office staff was hard hit during the war years. One man left in 1917 because he "found life at a mining camp entirely different from that of the city" and he "was completely at sea" where bookkeeping was concerned.[84] Later that year, another man left without warning to organize his own abstract company in Virginia City. He quit just as the annual reports were due, leaving Edmunds somewhat harassed and working at night. The war made things tight, Edmunds lamented, "and even a woman with any experience would be hard to secure."[85]

Edmunds also complained about another clerk, a man of little imagination and "not much force," who was the butt of numerous jokes around Ruby. "He is one of those fussy little Englishmen, neat and accurate in his books but fearfully slow, and if he gets out of his rut he is lost"—and, Edmunds added, certainly not worth the $137.50 a month he received. Yet given the wartime manpower shortage, the manager gave him a five dollar raise and did some serious thinking about hiring a woman, especially when he found one with nine years experience who was not too pretty and who seemed able to take care of herself.[86] Edmunds did hire a woman in his office that year, whether this one or another, but was not pleased with her short stay: "The only interest I saw her show was in five o'clock in the evening and pay day, so she had the opportunity of packing her trunk."[87]

High general wages and competition for labor disrupted Conrey's long era of good management-employee relations. One of the first inklings of discontent came when the three dredgemasters, Axtel Brynie, Olaf Pearson, and Olaf Kronwall, each with a dozen or more years of experience with the company, asked for an increase in salary to $200 a month—a raise of $20 for Kronwall and Pearson and $15 for Brynie. Brynie's five dollars a month extra went back to the time when No. 1 was so far up the gulch that he had to keep a horse and buggy to commute. All three had received bonuses of close to two hundred dollars each in December 1916. Sweeny and Edmunds tried to dissuade them from pressing their cause, pointing out that the dredges were on lean ground; but he passed their plea on to Jennings who promised to approach the directors. Jennings agreed with Edmunds that wage increases for the three would open the door to general raises, which was to be avoided. He thought perhaps more emphasis could be put on the bonus but also hoped that strong federal legislation would curb inflation.[88]

The dredgemasters made it clear to their employers that they were not making a demand, but a request. Several weeks later, one of the watchmen asked for shorter hours or more pay, and the operator of the steam traction engine threatened to quit if a raise was not forthcoming. Ruby officials refused the watchman and let the engine operator go but they recognized that to retain a good crew they might have to give increases across the board.[89] Edmunds and Sweeny calculated that based on average yardage and a $105,000 annual payroll, a 10 per cent raise would add a total of $875 a month or two-tenths of a cent per cubic yard, to operating expenses. They suggested, however, that there be no increase in the monthly or daily rates. The payroll should reflect the old wage after the employee's name, with the added notation "Plus 10%," which could be reduced or even eliminated if the high cost of living fell.[90] Once Boston ratified the plan, Jennings hoped the board's action would show employees the "desire of this company to extend them every justice" even when "struggling with a low-grade gold proposition."[91]

Meanwhile, Edmunds went east. When he returned to Ruby, he found that some of the outside workers had met to ask for an eight-hour day and time-and-a-half pay for overtime. Sweeny learned that the workers were about equally divided in the cause, so he simply fired five of the ringleaders and hoped that with winter coming the issue was at least temporarily settled.[92] With the labor situation at Ruby (and throughout the West) unstable, Hennen Jennings thought that it would be impolitic not to pay the employee bonus for 1917. He suggested, however, that because of smaller profits and recent wage increases the bonus be reduced from five thousand dollars to three thousand dollars, advice that the directors followed.[93]

But in 1918, another raise seemed warranted. The pressure, at least, was obvious early. With No. 2 closing down permanently, the company believed it had more breathing room but it had to fight to keep its workers from being drawn off.[94] Management at the dredges reiterated that those who left their jobs were gamblers, concerned more with short-term advantage than long-range possibilities. They lamented "the spirit of unrest, independence and lack of interest" that seemed to pervade the ranks of labor throughout the country, but they still began planning for another 5 per cent or 10 per cent raise, the decision to be left with the board in Boston. They also considered putting mechanics, linemen, and outside laborers on an eight-hour basis. Edmunds and Sweeny reported that those men who left did so in part because they could work a day shift elsewhere and have Sundays off; dredgemen at Ruby worked a seven-day work week and rotated regularly from one shift to another, spending two weeks on each.[95] Jennings passed to the board the manager's recommendations for new wages and hours, with his approval.[96]

Jennings supported a 10 per cent increase, but before any action was taken on the recommendation the crew on No. 4 struck with the support of the outside laborers. The occasion of the strike was a temporary shut-down of No. 4 because of low water. Manager Charles Kammerer had established the precedent that if a dredge had to be closed down for seven days or more, the dredge crew went on a nine-hour day, rather than eight, and took a reduction in wages. By 1918, only the change in hours was used in such circumstances, not the cut in pay. When it became apparent that No. 4 would be idle for several more days due to the drought, the company proposed the nine-hour shift, only to have the crew reject it. The employees argued that an eight-hour day was standard and that there was no reason to work more, especially because there were so many jobs to have elsewhere. One crewman told Sweeny he could make as much working six days a week somewhere else as he could in seven days at Ruby, and he would not have to change shifts regularly, which disrupted his family life.[97]

The actions of the dredge crew precipitated a strike of outside laborers, who held out for a raise of fifty cents per hour for either an eight- or a nine-hour day. Meanwhile, as Edmunds and Sweeny were putting together a new pay schedule proposal they received authorization from the Boston office to grant a 10 per cent wage increase. Their own plan, which the directors eventually adopted, was a series of compromises. The two managers proposed putting all workers on an eight-hour day, with general raises that varied from job to job. Outside laborers, for example, received the raise they asked for and were assigned eight-hour shifts, with the idea that they would have to "get into the collar" and do as much work as they had in nine hours; when more applicants appeared, the managers suggested, "we would weed out the drones."[98]

Boston approved the Edmunds-Sweeny plan. It met with general approval in Ruby as well, although there were one or two soreheads and one man quit in dissatisfaction. The pragmatic Edmunds thought it possible that the new plan "may probably put a little crimp in any further monkey-business" so far as labor agitation was concerned.[99] He continued to lament the high cost of labor and the difficulty of getting good men. He also noted that during peak years the company had employed about eighty-five workers, but in 1918 it had only about fifty-five on the payroll. The 1918 payroll would probably not reach $100,000 for the year, as contrasted with $116,000 for 1915, $111,000 for 1916, and $105,000 for 1917.[100]

Convinced that it was $2,500 well spent, the directors voted an employee bonus for 1918 and again the following year, but they warned that with dwindling earnings this might be the last.[101] In late 1919, the extra laborers hired to install additional gold-saving tables on No. 4 began to agitate for a raise of fifty cents per day, and when some of their fellow workers refused to sign a petition for more money they were thrown in the dredge pond. Blaming the IWW for the trouble, Edmunds and Sweeny "took the bull by the horns" and fired the agitators; two others left in sympathy.[102]

In September 1920, directors again authorized a $2,500 bonus or some other "way devoted to increasing the pay of the men employed in working the mine at Ruby."[103] Ten days later, twenty-five laborers signed a petition asking a for a fifty-cent raise for an eight-hour shift, but they received scant sympathy. The worker who had begun the campaign was fired. Hardly three months after voting the workers a bonus, the board predicted that "the Company has but a few months of life ahead" and suggested that a round of wage cuts might prolong its corporate existence.[104]

The last years of the Conrey operation were the most difficult from the standpoint of management's relations with labor. In 1921, President Charles Francis Adams II, who was also treasurer of Harvard College, bemoaned the situation when he wrote Frank Stanley, urging that a Boston officer be sent out to inspect the operation:

> I think we have all agreed for a long time that the quality of men operating our dredges was not such that we could leave them indefinitely without oversight. The results now being produced are not such that one can feel clear that everything is being done wisely and well.[105]

In retrospect, however, most observers of the Conrey Placer Mining Company would agree that labor-management relations had generally been good. Conrey's sporadic disputes were minimal when compared with the deep-level mine operators in Butte who had to contend with organized labor groups with a more radical outlook. Few other corporations paid regular bonuses to their employers, and few other mining firms enjoyed the easygoing and harmonious association among managers, workers, and villagers that existed in Ruby.

In many ways the town of Ruby and the Conrey Placer Mining Company were one. Named after the Ruby Mountains and located at an altitude of nearly a mile high, the small settlement was originally linked to the Madison County Poor Farm and before that to the chaotic early Alder Gulch gold fever. With the dredges, Ruby soon expanded to meet the demands of the company and received support from the firm in different forms. Under the watchful eye of Lawrence Brook, the town's first U.S. Post Office opened in 1901, coincident with the steam dredge era. Directors of the company in 1902 voted to subscribe one hundred dollars toward a schoolhouse and to lease land for the school at a nominal sum.[106] They also contributed one hundred dollars toward the building of "the fraternity hall at Ruby."[107] Although one visiting editor in 1906 predicted "thousands of people" in Ruby in the near future, even at peak the town's inhabitants numbered only in the hundreds. By 1909, Ruby had a population of some two hundred fifty people and was geared almost entirely to the Conrey dredge enterprises.[108] According to later sources, the town's size grew to about five hundred, and it boasted a post office, three hotels, three stores, a meat market, a church, a schoolhouse, and a town hall. In 1906, when Conrey built the power line for the dredges, Ruby became electrified. The brightly lit boats gave the town a festive air at night, and the company's monthly electric bill ran between ten and eleven thousand dollars.[109]

Most Conrey employees rented or bought property in Ruby, although some lived in Laurin or Alder. The 1910 census of Ruby Precinct showed fifty-four residents there who were connected with the Conrey dredging operations. Of these, forty-one were married, one owned a farm, and twenty-one owned their own homes.[110] The rest rented or, if single, perhaps boarded out with private families or at the Ruby Hotel, although Manager Edmunds thought "a fellow would be far better cooking for himself than eating at the hotel."[111]

Ruby was not a company town in any true sense, but Conrey did own housing. The housing came more or less by chance with the property it acquired for dredging, whether as part of the original Conrey Ranch, the County Poor Farm, or other land purchases. The Poor Farm property included the "Manager's" or "Director's" house plus two cabins. There was also a buggy shed, which had three-room living accommodations in it for the barnman. A dwelling west of the retort house, bought originally for five hundred dollars, was added to and remodeled for occupation by manager Edmunds. Two houses on the west side of Ruby, formerly occupied by Truman Prier and Dr. Townsend, had been bought through Judge William Clark; later two rooms were added to each. The Conrey house was a two-story, well-built structure. Houses also came with purchase of the Collins Ranch and the Beck property at Granite Creek, although the latter deteriorated quickly because it had no proper foundation and was situated on swampy ground. The company also owned two log cabins outside of town. The private business establishments were in Ruby, along with the company icehouse and a concrete tank that held an emergency water supply and that was used for swimming in the summer.[112]

The company provided these houses free or for reasonable rents to employees, most of whom were married and had children. When Charles Shott and his family moved into one of the Conrey residences in 1903, the editor of the *Times* reported: "The company has spared no expense to make its cottages comfortable, and a reservoir has been built which will supply them all with water."[113] Board and room in Ruby was not always easy to find, nor was it cheap, by Montana standards. In 1917, the cost of board increased from twenty-five to thirty dollars and then to thirty-six dollars a month, with the cost of a room increasing from seven to ten dollars a month.[114] Proprietors of the Ruby Hotel decided to close in June 1918, because at forty-six dollars a month for board and room, they were losing money.[115]

At times, the Conrey manager's house was used as a dining room for "unmarried officers of the Company."[116] At one point the concern had even empowered its acting manager to hire a cook and provide meals for employees in general—at the workers' cost—but this was not a policy for all of Conrey's life in Montana.[117] The company rented its south Ruby houses to workers at ten dollars a month for the two larger houses and eight dollars for the two smaller ones. Tenants paid forty cents per electric light per month, except that electricians and cleanup men who helped at the power substation and on the pole line and got their lights free.[118] In 1917, Edmunds let the new electrician from California have the log cabin close to the office but offered to open the "Managers House" and rent a room to an incoming engineer, although the house was then closed and the company would have to find a woman to take care of it and do the laundry.[119]

Although the company neither owned nor controlled most of the local businesses, aside from housing and boarding for some of its employees, it did assume an obligation for some semblance of medical care for isolated Ruby. It contracted with Dr. Townsend to provide medical services in town, paying him slightly more than two hundred dollars a month, at least during 1916 and 1917. Townsend resigned in 1917, apparently because his family did not like Ruby and his own health was not good; he relocated to Whitehall.[120] Hennen Jennings saw no benefit in attempting to dissuade Townsend from leaving and suggested a working agreement with a physician from Virginia City or Sheridan.[121] The company arranged with Dr. D. F. Clancy to visit from Virginia City from two-thirty to five-thirty each afternoon and at other times when he was called. Instead of a monthly contract like the one the company had with Dr. Townsend, each married employee contributed a dollar fifty and each single man a dollar a month; the company added an equal amount, thus saving itself from twenty to twenty-five dollars each month.[122]

This arrangement with Dr. Clancy was apparently still in force when the great influenza epidemic struck in the winter of 1918-1919. Like the rest of Montana, Ruby was hard hit. George Edmunds noted that at one time nineteen Conrey employees were ill with the flu, and "it practically demoralized our organization." Three men working on No. 1 dredge died, and dredge superintendent Sweeny was home with a 102 degree temperature. Dr. Clancy was in bed with the disease, and the company took drastic action, with Edmunds working full-time to obtain prompt medical service. The Conrey manager first appealed to the governor and the state Board of Health, who sent Dr. Arthur O. Jones of Butte to help out temporarily. Edmunds also enlisted the help of a former doctor-turned-rancher, and most important of all he was able to employ a nurse for the town. From "practically a panic," Mrs. A. O. Park of Dillon pulled the population through those trying days, Edmunds believed.[123] In the midst of all the misery came further affliction: "This is a very sad day all over Montana as the State goes bone dry at midnight," Edmunds wrote as 1918 came to a close.[124]

And of course the war touched both corporation and individuals. The town and the company were closely united in matters like the Liberty Loan drive. The Liberty Loan allotment was based on school district assessment, and Ruby's share was $8,750 in 1918, largely due to Conrey holdings. With only two dredges running and fewer employees on the payroll, George Edmunds was hard put: He hoped to raise $4,500, but deemed it an uphill struggle. Why not, he asked Secretary Stanley, get some of the New England stockholders to transfer part of their bond subscriptions to Montana?[125] In the end, a company contribution of $5,000 put the Liberty Loan drive over the top at Ruby, much to Edmunds's pleasure.[126]

When the war was over, Ruby celebrated. "Kaiser Bill and Little Willy" were burned in effigy, and everybody turned out for a spontaneous parade and gathered around a huge bonfire in the center of town to hear Edmunds "give a find patriotic address, interrupted many times by bursts of applause and enthusiasm." But for some, there was bitterness yet to come. Soon dredge superintendent O. O. Sweeny and his wife received a telegram from the Department of War that their son had died in France of pneumonia.[127]

Conrey men made the town go. Truman Prier, the Conrey engineer-draftsman-electrician, was elected superintendent of the Ruby Sunday School for 1906; ten years later, he was spending his spare time as business manager of the Ruby Hall company of theatrical players. A. H. Wiseman, who handled business matters for the company, cut a wide swath, appearing at numerous theatrical functions, speaking to the local schools on "Our Talents and How to Utilize Them," or serving as chairman of the Reception Committee for the opening of the new rebuilt KOTM Hall featuring Gorman's Columbian Gardens Orchestra.[128]

Local basketball teams, like the "Regulars" and the "Easter Eggs" (so named "because of their variegated costumes"), were sprinkled with Conrey workers and added spice to town life, especially at the dance after the big game. The Conrey nine practiced regularly and played good hardball and at night drew fifty couples dancing to the Westpahl Orchestra.[129]

Superintendent Charles Kammerer might at one time entertain President Elliott of the Northern Pacific Railroad; at another time, along with Truman Prier, he headed the committee in charge of the benefit supper and ball on behalf of little crippled Frankie Elphison.[130] On the Fourth of July, along with the normal oratory, horse races, baseball, and dancing, Conrey's dredges were opened to the public as part of the festivities.[131] The musical affair presented by Ruby's labor and dredge workers in early 1908, with plenty of refreshment, singing, and dancing, was labeled "the social event of the season."[132] Local editors recognized the relationship between the company and the town with frequent references to Ruby as "the town of many dredges" or "Marriage in Dredge Town."[133]

Ruby rejoiced when George H. Edmunds, then Conrey cashier, traveled to Indianapolis to marry Adeline Scott, daughter of a Hoosier businessman, and when the wife of Charles S. Baker, carpenter on one of the new dredges, presented her husband with a robust baby boy.[134] It beamed with pleasure when master machinist J. G. Crosser played in the local band.[135] It was noncommittal when electrician Le Roy Gilman ran as Democratic candidate for county treasurer—and was soundly trounced.[136] And it recoiled in horror at the passing of William Goodspeed of the Conrey machine shops who committed suicide by drinking a cocktail of whiskey and carbolic acid.[137]

Here was an interesting combination of small-town America—isolated, as it were—and of new industrial corporate technology. Ruby men were Conrey men. Ruby fortunes rose and fell with Conrey fortunes. Over the years, the operation showed a surprising degree of harmony between conservative management and a largely unorganized labor, in part because neither side took extreme positions and generally subscribed to conciliatory tactics.

Chapter Eight

A Mine Is a Wasting Asset

By the eve of World War I, the Conrey people were concerned with streamlining their operation. After a trip to Ruby in the summer of 1913, consulting engineer and vice-president Hennen Jennings made a careful study of projected future output and concluded that the Conrey and the Poor Farm companies should be joined into one.[1] Already, Conrey stockholders had voted to retire all of the $300,000 in First Preferred Shares.[2] In mid-July 1915, in response to Jennings's recommendation, they approved an increase in common stock by 2,500 shares of $100 each (to a total of 5,500 shares), raising capital nominally to $1,350,000, in order to acquire the property and assets of the Poor Farm Placer Mining Company, the shareholders of which agreed to sell in exchange to the new stock issue.[3] Part of the streamlining process was to whittle down capital and debt. Figures from Secretary Frank Stanley on October 21, 1915, indicated that the company had retired $2,000 of the Poor Farm common stock, $271,280 of Conrey First Preferred, $115,350 of Conrey Second Preferred, and $81,375 of Conrey common stock.[4] With assets calculated at $1,050,231.40 and a debt (apart from outstanding stock) of $26,829.40, the restructured concern was clearly a profitable venture.[5]

By this time, Charles Francis Adams II had become head of the company. A great-grandson of one American president and a great-great-grandson of another, Adams held but one share of Conrey stock in his own name, but he was Harvard's spokesman on the board. He was an active Boston lawyer, with many business interests of his own. As treasurer and a fellow of Harvard College, he was in charge of Harvard's capital funds for more than thirty years, and from 1929 to 1933 he served as Secretary of the Navy under Herbert Hoover.[6]

But any mine is a wasting asset. Mineral deposits are finite, and those of Alder Gulch were no exception. If the hard-rock adage that no miner could see farther ahead of his pick had some meaning, dredging offered at least a moderately better window into the future. Careful surveying and test drilling in advance of dredging operations could give an indication of what might be expected to a degree that was not possible in most other types of mining, but gold could not be added to the natural supply.

Thus, by 1917, Conrey people were talking about the beginning of the end at Ruby, planning as closely as possible how long active dredging could continue. Profitable ground was running out. The venerable electric dredge, No. 3, had already been honorably retired in October 1916, having worked for more than ten years and dug nearly 9.5 million cubic yards.[7] Informed observers like engineer Thomas H. Leggett, who visited Alder Gulch in the summer of 1917, recognized that there was not much gold left to mine.[8] Even so, engineers at Ruby spent a good deal of time drawing up a proposal to extensively renovate dredge No. 1. The renovation would include a new stern and spuds, major work on the hull, adding pontoons to increase flotation, and new and heavier equipment to enable replacement of the 7.5 cubic-foot bucket chain with the 9 cubic-foot chain of No. 3. The job would cost an estimated $34,560, a figure Hennen Jennings thought "certainly rather staggering," given the fact that very little undredged ground was still available.[9]

In 1917, a special agent of the Internal Revenue Service questioned the company's interpretation of the tax law and levied a back assessment of $6,469.38 for 1909-1915. Jennings protested bitterly, especially the application of the new war profits tax, arguing that the Conrey was a declining business with but a few years left and that the large depreciation on the property account was justified in order to cover losses by exhaustion. But a Supreme Court decision early in the year gave only a 5 per cent depletion allowance, which left Jennings puzzled and frustrated. When the Secretary of the Interior appointed Jennings chairman of a committee to study the gold industry in 1918, he determined to use the company's experience "to illustrate the injustice and folly of the excess war profits tax as applied to the gold industry." But the committee report was mild and brought no policy changes from Washington.[10]

When Jennings noted that operating costs for 1917 were 3.97 cents a cubic yard for No. 4 and 6.97 cents a yard for No. 2, he began to wonder if closing down the more inefficient dredge should not be seriously considered. He also thought it wise to assign Ambrose Burroughs, the engineer-draftsman on the scene, the task of sketching out plans for a new dredge that would utilize the good parts left from No. 2 and No. 3 when neither was running.[11] With the future somewhat indefinite, Jennings deferred renewal of the contract with the Montana Power Company, preferring to wait until expiration of the old one on March 1, 1919.[12] In April 1918, Edmunds predicted that No. 2 would be closed down permanently in a month or so and that Secretary Stanley should adjust the insurance accordingly.[13]

No. 4 was doing remarkably well in 1918, reworking the *Maggie Gibson* tailings: from October 23, 1917, to February 7, 1918, No. 4 took out $93,003.87, or 8.87 cents a cubic yard, for a profit of 5.02 cents a yard.[14] But No. 1 and No. 2 were doing poorly. No. 2 would soon be finished. Edmunds wrote Jennings early in May that

> No. 2 dredge finished excavating virgin ground yesterday, and is digging a shallow pond for its last resting place. One thing can be said about this dredge, in the last several weeks it has been dredging some of the tightest ground we have, it has put up a good fight, and it has demanded careful nursing by frequently demanding exceptional minor repairs.[15]

A few days later Edmunds wrote again: "No. 2 dredge finished digging yesterday, is now safe in its 'watery grave,' and we are taking the bucket chain off to transfer it to No. 1 dredge." Long-time employees, like Olaf Kronwall, the dredgemaster, were honored with food, drink and "a beautiful electric percolator" and found jobs elsewhere.[16] No. 1 and No. 4 continued. No. 4 had been "the salvation of this property so far as dividends have been concerned. . . ." It would "die with a wonderful reputation and the admiration of every one who has seen it,"

Edmunds intoned. "It will not die from degeneration but from starvation—not enough nutriment in what it has to eat."[17]

No. 1 and No. 4 ran through all of 1918; No. 2 ran only about four months. The Conrey Company, employing an average of fifty-five men dug a little more than 3.35 million cubic yards that year.[18] But the margin of profit continued to shrink, and Jennings was convinced "that the continuance of work at Ruby is very problematic. . . ."[19] But the work went on. The company involved itself in a rare legal dispute over the diversion of water from Mill Creek.[20] When a California dredge specialist visited Ruby in May 1919 and reported that California dredges carried half again as much gold-saving area as No. 4, the directors authorized the expenditure of up to $10,000 to add another deck of gold-saving tables.[21] Bad weather in November and early December hampered the dredges, and the clean-up on No. 1 was "quite depressing," without much hope for profits.[22] Thanks to a warming Chinook wind and some dynamite, No. 1 was soon started up again, but the prognosis was not good.[23] When the directors voted the usual employees Christmas bonus at the end of 1919, they warned that there might not be another.[24] Yet, there was. In September 1920, after two of the directors had visited Ruby, the board again sanctioned a bonus and at the same time authorized the payment of fifteen dollars a share to partially retire the capital stock.[25] All the while, officials were aware that operations were winding down. It was "similar to any deathbed scene," Edmunds remarked, "that while we looked forward to the inevitable, yet when death comes it is a shock."[26] Secretary Stanley was authorized to work with Edmunds in closing out operations when the time came. He was to reiterate the fiscal bleakness of the picture and inform him that "the Company has but a few months of life ahead, with practically little promise for the shareholders."[27]

The company was still paying dividends in 1921. Ordinarily, tax protests were wasted effort, but that year George Edmunds succeeded in getting the county net proceeds tax cut from $2,336.11 to $1,174.80 and the state license tax reduced from $782.02 to $383.76, with further hope that the federal tax might also go down.[28] Gone were the halcyon days. Any small charity was welcome.

By this time, Secretary Frank Stanley had died, having held his position for most of Conrey's corporate life. The board authorized the new secretary-treasurer and the president, Charles Francis Adams II, to sell the real estate and water rights and, after dredge No. 4 finished up and reached the Alder railroad station, to dispose of other assets. By mid-year, insurance on No. 2 and No. 3 had been cancelled, although the company still carried $100,000 on No. 4.[29] In July 1922, directors empowered the treasurer to sell the company's securities from time to time to meet obligations and they voted a 10 per cent dividend in liquidation. Four months later, on November 14, the directors voted another dividend (12 per cent) in liquidation and authorized the president and treasurer to proceed with dissolution of the company, a process already underway.[30] But winding up took time. It was not until March 1927 that the capital stock was completely retired and final payment of dividends in liquidation was made.[31]

Part of the winding down process was disposing of the assets, dredges, real estate, and other tangible property. In 1918, Edmunds and Sweeny gave rough estimates of what the plant might bring on the market: No. 1, $7,800; No. 2 and No. 3, $10,000 together; No. 4, $25,000; machine shop and tools, $7,000.[32] This was only a fragment of cost, and Jennings thought the managers at Ruby were "over-conservative" in their appraisals.[33] Jennings had earlier suggested advertising dredge No. 2 in the mining periodicals, but now he changed his mind and believed the best approach was to work through a middle-man specialist on a commission basis. He was thinking of Charles Janin, second generation in a large and important family of mining engineers, well known in the West since the 1860s. Janin foresaw a revival of dredging in Russia and believed the Soviet Union to be a budding market for both new and used dredge machinery.[34] Conrey inventories near the end of 1918 showed $105,330.33 on hand in machinery, equipment, and supplies, apart from the dredges themselves.[35]

Early in 1918, Albert Hanford, who knew Jennings and who was interested in the large Natomas Consolidated dredging operations in California, inquired about purchasing No. 3 and was sent information, including a copy of Jennings's *History and Development of Gold Dredging in Montana*, but there was no follow-up.[36]

Apparently at Charles Janin's instigation, representatives of the Columbia Steel Company of San Francisco visited Ruby to inspect No. 2 and No. 3 and concluded that it would not pay to ship them very far. The bucket chain on No. 2 had only two years left, they pointed out, and both dredges had been bypassed by the great improvements made in the field over the past decade. Conrey would be fortunate to get $20,000 for the two of them, the Californians predicted.[37] Janin urged the directors to be patient. Above all, he cautioned, the dredge machinery should not be sold as scrap; if nothing else, it would be valuable as replacement parts, either in the United States or abroad.[38]

Charles Janin scurried around, using his excellent contacts in Europe and Asia. He had inquiries from Russian firms and was aware that the Chinese were contemplating dredge operations in Manchuria. He thought that the best parts of the Conrey equipment might be used in one of those enterprises.[39] Janin informed the Conrey people that the price was a matter of negotiation and that his commission would be 10 per cent of the selling price.[40] In July 1919, the directors gave him exclusive rights for the rest of the year to sell No. 2 and No. 3.[41]

But the outlook was not bright. When Oroville engineer George Carr visited Alder Gulch in the spring of 1919, he estimated that No. 2 and No. 3 were from 50 to 60 per cent worn out and observed that on the market they were competing with a number of complete dredges in California that were in much better shape. He counseled waiting, but thought that eventually Conrey might have to sell for scrap—and even the scrap market was in a slump.[42] In the meantime, Janin had a nibble from Mexico, a Missoula group evidenced some interest, and it looked for a time as if A. H. Cox & Company of Seattle might take some $15,000 of the equipment but ended up buying only a small amount.[43] Edmunds thought he had a chance to sell the No. 3 bucket chain to John G. Barker of California. Barker, he said, "thinks he is a mighty big guy" who might not be aware of George Carr's adage: "Only a poor concern or a fool would buy a secondhand line."[44] Precisely what equipment Baker purchased is not clear, but he gave four 6 per cent notes on La Grange Gold Dredging Company, of which he was president.[45]

At the end of 1920, the Conrey Company was well into dismantling No. 1 dredge, but it had not yet begun on No. 2 and No. 3.[46] Nine months later, when the directors sent Director Walter Bigelow out to Montana, they announced they would be willing to take shares of the purchasing firms in exchange for their machinery.[47] At one point, Charles Janin believed that the Lena Gold Mining Company in Russia was about to resume operations and might be interested in No. 4, once its work was finished. In fact, Janin suggested that the Lena's English backers purchase what was left of the Conrey operation "as a sort of school of dredging experience" before moving and reinstalling the big dredge in eastern Russia. George Edmunds thought this idea was promising, "because very likely conditions here are quite similar to what they would encounter in Siberia."[48] But nothing came of the suggestion. For a time it looked as if the Swank Creek Mining and Dredging Company might take one of the dredges for its Washington property, although it was short on funds.[49] Now back onto the scene came A. H. Cox & Company of Seattle, asking a thirty-day option to buy. Cox's David H. King, who visited Ruby with the manager of the Swank Creek Company (a "babe in the woods") in mid-July 1921, had represented the dredge as "ninety-seven percent new," when actually it had been worked for nearly twelve years. As Edmunds put the story together, Cox & Company was convinced that the Swank Creek ground was rich and "by hook or by crook" wanted in, hoping to take over the property if an unsuitable dredge were put on it and did not pay immediately.[50]

Only the machinery was saleable, of course. Hulls were cheaper to build new than they were to move any distance. At the beginning of August 1921, Edmunds reported that No. 1 was completely dismantled, that No. 2 was practically taken apart, No. 3 was being torn down, and No. 4 was still operating. But Edmunds reported that much of the equipment was not of standard design and that dredges were a glut on the market.[51] Subsequently, No. 1 and No. 2 were sold—at least their machinery—to the Pioneer Gold Dredging Company of Baker, Oregon, and were eventually scrapped.[52] The final disposition of No. 3 is not clear: it was dismantled, and perhaps the La Grange Gold Dredging Company took part of it. Number 4 closed down for the last time in the spring of 1922. By mid-June, clean-up man Leslie Gilman was one of the few remaining company employees. The others had scattered: oiler Joe Buford was in Missoula; dredgemaster Olaf Kronwall had a new job in Butte; Axtel Brynie, in charge of the machine shop, had moved to Portland; superintendent Sweeny was in southern California, where he hoped to recapture his health. In 1924, Washington recognized the inevitable and officially closed the Ruby Post Office, an institution whose lifespan roughly matched that of the dredges themselves. Marion Steam Shovel Company records indicate that No. 4 was not sold intact but was scrapped by the Conrey Placer Mining Company, probably shortly after it shut down in 1922 and sold for the metal to Northwest Steel & Metal Products of Seattle.[53]

The real estate also was to be disposed of. No one would pretend that dredged-over ground would sustain agriculture, although there was some evidence that it might be useful for grazing. In 1918, the company allowed its employees to fence in parts of the land that had been worked by No. 3 dredge and wrote the Agricultural Experiment Station in Bozeman about planting sweet clover seed on the tailing piles, an experiment that the experiment station encouraged.[54] But even without reclamation, the ground dredged by the earlier flume-type dredges, with their low and practically flat tailing piles, might prove useful as pasture. As early as 1913, on behalf of the company, Charles Kammerer had leased land on the old Collins Ranch to local ranchers; two years later, he followed suit with the Elling Ranch land. In 1918, the Anderson brothers negotiated to lease the Collins land and buy the Elling property (240 acres) at seventy-five dollars an acre—fifteen dollars more than Conrey had paid for it—but no deal was completed until early 1919.[55]

Most of the real estate went to individuals in small lots, with a flurry of transactions between 1920 and 1923, although there was still transfer by purchase at tax sale as late as 1939.[56] An occasional sale involved seventy or more acres, and in 1922 the company turned over 305 acres to Madison County, but Conrey still held over 832 acres when it began liquidation proceedings in November and put its property in the name of President Charles Francis Adams.[57] Most transfers involved two to half a dozen acres and often were to names once on company payrolls.[58] Northwest Steel & Metal Products of Seattle bought the machine and blacksmith shop, and Leslie Gilman, the veteran clean-up man, acquired the office, retort house, and adjoining grounds, the setting for the Gilman family home down to recent times.[59]

If the company was remarkably free of litigation, unlike the rich deep-level mines on the Comstock or at Butte, this was a function of dredging in general, not Conrey in particular. Placer grounds were generally clear-cut in definition: there were no lodes, with their "dips, spurs and angles" to follow; there was no perplexing application of the "apex law," which enabled adjacent owners to challenge a successful concern. Hence, Conrey made few trips to the courtroom, and when it did it was over relatively trivial matters, as viewed through the eyes of the mineral world.

During its twenty-four years in Alder Gulch, the Conrey Placer Mining Company proved to be both innovative and conservative. Its bold ventures with the aerial tramway and the Hayward "orange peel" buckets were novel attempts to adapt technology from other industries to placer mining. Its early commitment to the bucket-line dredge and its willingness to experiment with machines in their infancy in America made it one of the pioneers of the new technology. At the same time, Conrey funding was essentially conservative. It was "courageously planned and wholesomely financed," as Hennen Jennings phrased it. A small corps of investors began the enterprise and was heavily dependent on capital provided by Gordon McKay, who held the controlling interest at the time of his death.[60] When capital was reorganized and expanded, it was essentially the members of this same well-knit group who extended their holdings. The final statement of retirement of capital stock in 1927 lists twenty-two shareholders, most of them descendants of or trustees for the original investors. Among them are the estates of McKay, Bigelow, Jennings, Clark, and Stanley, plus Walter Bigelow, Page Shaler Page (a grandchild of Nathaniel Shaler), and Coleman Jennings.[61] There had been no public offerings, no speculation in company issues. New equipment had been raised essentially by loans from original investors in the form of preferred shares and had been paid off out of operating profits.

During the steam era, prior to 1906, the cableway, "orange peel" buckets, and the three dredges (including the *Maggie Gibson*) worked a total of 6,400,290 yards and extracted $1,469,296 in gold bullion—an average of 22.96 cents per cubic yard. During the period of electric-powered dredging, from 1906 to 1920, the four newer dredges dug 51,861,111 cubic yards, from which

Leslie Gilman retorted his forty-five-pound bars worth $6,780,240 from fine gold that ran 13.08 cents a yard. The grand total, by company account, was 58,261,401 cubic yards of gravel processed and $8,255,536 in gold, for an overall average of 14.14 cents a cubic yard.[62] No. 4, at least, worked another year or so, but the total is somewhat short of outside estimates of $13,000,000 over Conrey's corporate life.[63]

The year 1915 was the company's best season, when it had a gold return of $856,937 and paid dividends of $184,080 or about 57 per cent on its common stock. It was either that year or the previous one when Conrey was considered important enough that a film crew came to Ruby to make a movie of the dredges, including the pouring of gold bricks, to be shown at the Panama Pacific Exposition in San Francisco by the United States Bureau of Mines.[64]

Gold in hand is one thing; profit is another. Conrey records are sketchy for the early years, but they indicate that the company did not make much money, if any, until the bugs were worked out of the steam dredges. And even the early electrics caused problems and required additional investment for rebuilding. But eventually, especially after the three California-type dredges, No. 1, No. 2, and No. 4 electric, were on a stable footing, Conrey became a paying proposition. Records show that thirty-three dividends were paid on regular stock between October 1913 and March 1920, plus five others on preferred shares for a total of $909,628.79. In addition, during the winding down operations that ended in 1927, shareholders received another $222,360 from land sales and retirement of capital stock.[65] Near the end of its corporate life, *The Mines Handbook* described the company as having been "one of the most profitable dredging enterprises in America. . . ."[66] Precisely how much Conrey gold went to Harvard University is difficult to say, mixed as it was with Gordon McKay's many other assets and accumulating year by year after his death in 1903. By the time the entire estate came into the university's hands by the middle of the twentieth century, its net worth was estimated at well over $15 million—perhaps even $20 million, of which between $1.5 million and $2 million was probably Conrey profits.[67]

Where does the Conrey Placer Mining Company fit into the broader scheme of dredging and placer mining in Montana?

To begin with, it was the premier dredging concern in the Treasure State. Although earlier, the Grasshopper Creek dredges were of fleeting existence; and for Conrey's two score and four years, its machines were the only ones in continuous operation. Other dredges came and went—at Bannack, French Gulch, Dry Cottonwood Creek, Hudson, or Helena—but only the Ruby machines remained constant year after year. Between the floating of the *Fielding L. Graves* in 1897 and the retirement of Conrey No. 4 in 1922, there were at least sixteen other dredges built in Montana, but most lasted a year or two at most and sometimes only a few weeks at the least. As of 1916, it was estimated that in terms of yardage, Conrey had been responsible for about 80 per cent of the dredging done in Montana.[68]

At the same time, it should be pointed out that while placer deposits were important, they were not the only source of gold in Montana. In 1911, for example, the state took out $684,801 of the precious yellow metal from placer ground, of which Conrey's Madison County dredges produced $584,948. But deep-level mines produced $3,025,770 that year, $275,441 in Madison County.[69] For Montana in general, the amount of gold dredged was probably not more than one-quarter of the total, even in peak years, unlike in California or Alaska, where the percentage was a great deal higher. But among placer methods, the production of bucket line dredges far surpassed dragline, dry-land, hydraulic or small-scale hand operations.

What of Alder Gulch itself? After the last Conrey dredge closed down, that part of Madison County experienced a dozen lean years, until the Humphreys brothers of Denver took an interest during the Depression. The sons of an adventurous pioneer entrepreneur, Albert E. and Ira Boyd Humphreys already had investments in minerals and oil in various parts of the West, especially in Colorado, California, New Mexico, and Arizona, before they took options in Alder Gulch late in 1934. Then they were "all busy as bird dogs," as Albert put it, on the design of the new 5,000-yard dry-land dredge that started up early the next spring. With the aid of a quarter of a million dollar loan from the Reconstruction Finance Company, Humphreys Gold Corporation's combination shovel and dragline—described as "a whale of an affair"—took out 19,591.7 ounces of gold worth more than $685,700 before it stopped work in June 1937 and moved its equipment to California.[70] Another dry-land dredge worked part of the area briefly in 1940-1941, recovering some 1,180 ounces, but by this time the gulch deposits were pretty well exhausted. But these were not boom years. At the end of the 1930s, if not slumbering, Ruby was nodding. About thirty people occupied the old buildings and kept the frame store and run-down gasoline stations in business. During World War II, gold mining was curtailed as nonessential, and defense industries on the Pacific Coast drained Ruby and Virginia City even more. After three-quarters of a century, and having been worked with nearly every conceivable gold-saving device, Alder Gulch was played out.[71]

The dredges left the terrain ripped inside out. They drastically rearranged the landscape, in the process erasing earlier camps—Nevada, Central, Junction, and Adobetown. What had once been stockland or farmland along a pleasant stream became more than a dozen miles of windrowed rocks, thirty to forty feet high, with a few dredge ponds stocked with trout. Half a century later, as shown by Bureau of Land Management before-and-after photographs, a few hardy species of plant life would soften the terrain. Their tenacious roots struggling into the thin soil trapped amid once naked gravel and boulders, tough timothy, wheatgrass, red clover, wild rose, sagebrush, cottonwood, and clumps of alders provide at least limited relief.[72] By California and Alaska standards, it was not a large area—four hundred to five hundred acres; a good deal of it had previously been devastated by earlier miners, and it was tucked away out of sight of the Interstate.

But contemporary Montanans made few complaints. If Olin D. Wheeler, a spokesman for railroad tourism, in 1902 mentioned a dredge "scooping up a fine alfalfa field" at the mouth of Alder Gulch, he wrote more with an acceptance of economic inevitability than as a lament.[73] Conrey was a vibrant industry. It paid out $42,928.16 for power alone in 1920, not one of its peak operating years.[74] It provided jobs and was a leading taxpayer in Madison County, generally ranking behind only the Northern Pacific Railroad Company, the Madison River Power Company, and the Oregon Shortline Railroad.[75] Citizens in the early twentieth century were not concerned with despoliation of the environment and Montana newspapers are remarkably free of any condemnation. In California during the Progressive era, the California Federation of Women's Clubs spearheaded a drive to curb dredge pollution and some farm spokesmen joined in to bring the introduction of a number of bills in the legislature to require soil to be left "tillable for agricultural purposes." But despite such rumblings of protest, no concrete responses had been forthcoming.[76] In Montana, during the Conrey era, not even these limited undertones of complaint were apparent. The ravishing of the landscape was part of the price of progress, part of what Americans expected to pay for the exploitation of natural resources, whether in the coal regions of Pennsylvania, the oil fields of Texas, or the placer gulches of Montana.

Notes

Chapter One: Introduction

1. Part of this chapter appeared originally in Clark C. Spence, "The Golden Age of Dredging: The Development of an Industry and Its Environmental Impact," *Western Historical Quarterly* 11 (October 1980): 401-414.

2. Arthur Lakes, Jr., "Gold Dredging Practice in Placers of Breckenridge, Colorado," *Mining Science* 59 (January 14, 1909): 28. For a more elaborate but comparable definition, see D'Arcy Weatherbe, *Dredging for Gold in California* (San Francisco: Mining and Scientific Press, 1907), p. 46.

3. The dragline dredge often consisted of a gold recovery plant floating on a hull in a pond, but without a digging chain, that was fed by a dragline shovel or shovels on the bank. The dry-land dredge was also fed by a power shovel, generally of a dragline sort, but its washing plant was mounted on skids or wheels. Both versions were much less expensive than the bucket ladder type and were much less efficient. Although suction dredges have been tried many times in gold-bearing areas the world over, rarely have they been used profitably. Charles J. Lyden, *The Gold Placers of Montana,* Memoir No. 26 (Butte: Montana Bureau of Mines, 1948), p. 5; Walter R. Crane, *Gold and Silver* (New York: Wiley, 1908): 393-394.

4. John Gunther, *Inside U.S.A.* (New York and London: Harper & Row, 1947), p. 159; *Hardin Tribune* (Montana), October 21, 1921.

5. *Alder Gulch Times* (Montana), June 16, 1899; George E. Walsh, "Modern Gold Ship," *Scientific American Supplement* 64 (July 6, 1907): 3-4; "Flagships of the Gold Fleet," *Popular Mechanics* 69 (May 1938): 730-733; Charles S. Aiken, "Farming for Gold," *Sunset* 23 (December 1909): 651-656; *Mining and Scientific Press* [MSP] 97 (November 28, 1908): 734.

6. Anna Morrison Reid, "The Gold Dredging Era," *Overland Monthly* 39 (June 1902): 9921; Nome Gold Dredging and Power Company, *The Gold Ship* (New York, c. 1907), p. 5; Franklin R. Carpenter, "The History and Production of Gold," *Report of the Director of the Mint Upon the Production of Precious Metals in the United States,* 59th Cong., 1st sess., 1904, H. Doc. 19 (Serial 4981), p. 52; *Red Book Magazine* 11 (August 1908): n.p.; Walsh, "Modern Gold Ship," *Scientific American Supplement* 64 (July 6, 1907): 4.

7. J. Wesley Jones, "Jones' Pantoscope of California," *California Historical Society Quarterly* 6 (September 1927): 244; Earl Ramey, "The Beginnings of Marysville," *California Historical Society Quarterly* 15 (March 1936): 51-52; Gilbert Chinard, "When the French Came to California," *California Historical Society Quarterly* 22 (December 1943): 309; Aiken, "Farming for Gold," pp. 652-653; Almarin B. Paul, "How Stage Catastrophes Are Brought About," *San Francisco Chronicle,* August 4, 1907.

8. *Engineering and Mining Journal* [EMJ] 14 (August 27, 1872): 138; 35 (June 23, 1883): 360; 42 (September 25, 1886): 228; MSP 45 (July 8, 1882): 24; *San Francisco Bulletin,* November 11, 1878; *Engineer of the Pacific* 1 (June 1878): 2, 3; 2 (June 30, 1879): 8; *Helena Daily Herald* (Montana), March 1, 1879, August 4, 1880, April 7, 1886; Anna Morrison Reid, "The Dredging Era," *Overland Monthly* 39 (June 1902): 990.

9. R. S. M. Sinclair, *Kawarau Gold* (Dunedin, New Zealand: Whitcombe & Tombes, Ltd., 1962), p. 14.

10. Province of Otago, *Votes and Proceedings of the Provincial Council,* Session 21, 1865-1866 (Dunedin, New Zealand: Daniel Campbell, 1866), p. 42A; *Dunstan Times* (New Zealand), April 30, 1869.

11. Thomas A. Rickard, "Alluvial Mining in Otago," *Transactions of the American Institute of Mining Engineers* 21 (1892-1893): 464; EMJ 70 (August 4, 1900): 129; "Appendix to Report on the Gold Fields of New Zealand," *Appendix to the Journals of the House of Representatives of New Zealand,* vol. 2 (Wellington, 1876), p. 8.

12. EMJ 70 (August 4, 1900): 129; Charles Janin, *Gold Dredging in the United States,* Bulletin No. 127 (Washington, D.C.: U.S. Bureau of Mines, 1918), pp. 5-6; *Dunstan Times,* June 30, July 21, September 1, 1882; *A Plea for Protection: Together with an Account of the Industries of the South Island* (Dunedin, New Zealand: Fergusson & Mitchell, 1888), pp. 102-105; E. J. Von Dadelszen, comp., *The New Zealand Official Year-Book, 1899* (Wellington: John McKay, Government Printer, 1899), p. 509.

13. "The Mining Industry," C-2, *Appendix to the Journals of the House of Representatives of New Zealand,* vol. 1 (Wellington, 1889), p. 145.

14. Vivian Maureen Cartmell, "Pactolian Sands of Otago: An Historical Geography of Gold Dredging in the Otago Mining District" (M.A. thesis, University of Otago, New Zealand, 1969), pp. 4-5.

15. Robert McIntosh, *Three Prize Essays on the Present Condition and Future Prospects of the Mineral Resources of New Zealand, and the Best Means of Fostering Their Development* (Wellington: John McKay, Government Printer, 1907), p. 92.

16. EMJ 70 (August 4, 1900): 129-130.

17. *Who's Who in Engineering, 1922-23* (Brooklyn, New York: Lewis Historical Company, 1922), p. 1009; *Who Was Who in America, 1961-1968,* vol. 4 (Chicago: Marquis, 1968), p. 761; *The Cyclopedia of New Zealand,* vol. 3 (Christchurch: Cyclopedia Company, Ltd., 1903), p. 873; MSP 109 (September 19, 1914): 432. In 1910, Risdon was bought out by the Union Iron Works, which in turn was absorbed by Bethlehem Steel. Under one name or another, the company has continued to manufacture mining equipment in San Francisco, dredges included, down to recent times.

18. Charles M. Romanowitz, "California's Gold Dredges," *Mineral Information Service,* California Division of Mines and Geology, 23 (August 1970): 165; "Dredging for Gold in the Western States," *Overland Monthly* 42 (September 1903): 265; George L. Holmes, "Placer Dredging in California," *Mining World* 26 (May 25, 1907): 654; Frank W. Griffin, "Reminiscences," *Western Mining News* 3 (June 1935): 12. It is ironic that local newspapers later gave Hammon the credit for initiating dredging in California. Prior to his time, one editor wrote that "an attempt at dredging had been made in a crude manner in

New Zealand." Undated clipping from *Oroville Daily Mercury,* in Natomas Company and Its Predecessors, 1851-1956, II, pp. 898-899, California State Library, Sacramento.

19. Lyden, *Gold Placers of Montana,* Plate 1, caption; *Dillon Examiner,* April 10, May 22, 1895. Some writers, including Lyden, gave the date as 1894. See EMJ 78 (August 4, 1904): 174; Randall E. Rohe, "Gold Dredging in the American West: Origin and Diffusion," *The Pacific Historian* 28 (Summer 1984): 6. Contemporary newspaper stories leave no doubt as to the chronology.

20. *Dillon Examiner,* October 2, 1901; MSP 71 (November 30, 1895): 353; Norman Cleaveland, "Bucket Line Dredge Methods," *Mining Congress Journal* 27 (November 1941): 19.

21. John Van Jon, "There's Gold in That Gravel," *Excavating Engineer* (December 1938), pp. 606-607, Charles Romanowitz MSS, University of Alaska Archives, Fairbanks [Romanowitz MSS]; *Dillon Examiner,* March 20, April 10, May 22, 1895. Harper wrote one editor on November 29, 1897: "I designated the ground, drew up the original plans, built and launched the dredge, was the original promoter of the company and its manager during the period of the construction of the plant. Mr. Reiling, who knew nothing whatever of placer mining or dredges, was connected with the enterprise simply as one of the promoters of the company and its president." See EMJ 64 (December 4, 1897): 663.

22. *Dillon Examiner,* October 17, 1897; Robert Norman Bell, "Dredging for Gold," *Mines and Minerals* 19 (March 1899): 380; W. S. Russell, "Gold Dredging," *Mines and Minerals* 21 (December 1900): 196.

23. Hennen Jennings, *The History and Development of Gold Dredging in Montana,* Bulletin No. 121 (Washington, D.C.: Bureau of Mines, 1916), pp. 5-6.

24. Romanowitz, "California's Gold Dredges," p. 165; Weatherbe, *Dredging for Gold in California,* pp. 65, 66, 69; Lewis E. Aubury, *Gold Dredging in California,* California State Mining Bureau Bulletin No. 57 (Sacramento: W. W. Shannon, 1910), pp. 38, 40, 41; Griffin, "Reminiscences," 14; *Mining World* 24 (May 19, 1906): 607-608; *Mining and Metallurgy* 25 (May 1903): 461; Norman Cleaveland, "Some Comments on Dredging as a Mining Method," *1967 Proceedings of Wodcon, World Dredging Conference* (n.p.: World Dredging Conference, 1967), pp. 13.

25. H. H. Dunn, "Eight-Cent Gold-Digger Earns Millions," *Popular Mechanics* 49 (January 1928): 77; Hennen Jennings, "Report on the Conrey Placer Mining Company's Property, Ruby, Montana," 1906, vol. 14, Conrey Placer Mining Company MSS, Baker Library, Harvard University [Conrey Papers]; EMJ 90 (July 30, 1910): 202; 91 (March 4, 1911): 483; Aubury, *Gold Dredging in California,* pp. 54, 55-56, 57; Charles Romanowitz, "The Uses of Floating Dredges for Mining" (paper presented before the National Western Mining Conference of the Colorado Mining Association, Denver, February 4, 1966), Romanowitz MSS.

26. Weatherbe, *Dredging for Gold in California,* p. 74; George Walsh, "Economy in Mining with the Modern Gold Dredge," *Mining World* 24 (May 19, 1906): 608; Charles Janin, "Gold Dredging on the Seward Peninsula," MSP 105 (September 28, 1912): 398; Janin, "Gold Dredging in Alaska and the Yukon," *Mining Magazine* 6 (January 1912): 45.

27. Mary Lee Davis, *Uncle Sam's Attic: The Intimate Story of Alaska* (Boston: W. A. Wilde Company, 1930), pp. 141-142; Philip S. Smith, *Mining Industry of Alaska in 1934,* Bulletin No. 868A (Washington, D.C.: U.S. Geological Survey, 1936), p. 58; MSP 102 (January 7, 1911): 20; 108 (January 10, 1914): 143; 124 (February 4, 1922): 147; *Mining and Metallurgy* 14 (May 1933): 215.

28. Janin, *Gold Dredging in the United States,* p. 11; MSP 112 (March 4, 1916): 346; *Mineral Resources of the United States, 1911* (Washington, D.C.: Government Printing Office, 1912), pt. 1, p. 462; *Mineral Resources of the United States in 1920* (Washington, D.C.: Government Printing Office, 1921), pt. 1, p. 163; Philip S. Smith, *Mineral Industry of Alaska in 1940,* Bulletin No. 933A (Washington, D.C.: U.S. Geological Survey, 1942), p. 12; *Minerals Yearbook, Area Reports, 1953,* vol. 3 (Washington, D.C.: Government Printing Office, 1956), p. 74; *Minerals Yearbook, Area Reports, 1960,* vol. 3 (Washington, D.C.: Government Printing Office, 1961), p. 78.

29. EMJ 85 (May 30, 1908): 1086; 92 (September 30, 1911): 636-640; W. H. Lanagan and R. G. Smith, "Dredge Data from Oroville and Folsom Districts," February 1911 (typescript), Fred L. Morris MSS, Carton I, Bancroft Library, University of California, Berkeley.

30. Charles Janin to Edgar Rickard, May 19, 1916, Janin to George W. King, December 30, 1916, Janin to Chester Purington, February 21, 1919 (copy), Charles Janin MSS, Box 6, Huntington Library, San Marino, California [Janin Papers]; "Dredging Operations of the World," *Mining and Engineering World* 44 (January 1, 1916): 41-44.

31. "Flagships of the Gold Fleet," *Popular Mechanics* 69 (May 1938): 730-731.

32. *Fourteenth Census of the United States taken in the Year 1920: Vol. XI, Mines and Quarries* (Washington, D.C.: Government Printing Office, 1922), p. 379; Henry H. Symons, "California Mineral Production and Directory to Mineral Producers for 1932," *Bulletin,* No. 109 (Sacramento: California Division of Mines, 1933), p. 36; MSP 117 (November 23, December 21, 1918): 689-690, 837; 124 (January 14, 1922): 55; *Commissioner of Mines Report for the Territory of Alaska, 1941-1944* (Washington, D.C.: Government Printing Office, 1945), pp. 16-18, 26; Lyden, *Gold Placers of Montana,* p. 39; *Mines Magazine* 24 (July 1944): 339; *Minerals Yearbook, 1964, Area Reports,* vol. 3 (Washington, D.C.: Government Printing Office, 1966), p. 94.

33. *The Madisonian* (Virginia City, Montana), January 27, 1922.

34. EMJ 83 (June 29, 1907): 1225.

35. EMJ 84 (July 13, 1907): 69.

36. Jennings, *History of Dredging,* p. 21.

37. EMJ 84 (July 13, 1907): 69; MSP 100 (March 19, 1910): 431.

38. Hennen Jennings to George Edmunds, July 29, 1917 (copy), Edmunds to Jennings, August 11, 1917, Hennen Jennings MSS, American Heritage Center, University of Wyoming, Laramie [Jennings MSS].

39. Jennings to Edmunds, February 1, 1918 (copy), Jennings MSS.

40. *Mining and Engineering World* 44 (January 1, 1916): 41-44.

41. George J. Carr to Hennen Jennings, May 26, 1919, vol. 30, Conrey Papers.

Chapter Two: Of Men and Money

1. Robert Raymer and Tom Stout are among the few historians to mention the Conrey Placer Mining Company. Raymer devoted one paragraph of his

three volumes to the firm and its activities. Stout provided two short sentences. See Robert George Raymer, *Montana: The Land and the People,* vol. 1 (Chicago and New York: Lewis Publishing Company, 1930), p. 595; Tom Stout, *Montana: Its Story and Biography,* vol. 1 (Chicago and New York: American Historical Society, 1921), p. 771.

2. EMJ 83 (July 29, 1907): 1223, 1225; *Mining Science* 59 (March 4, 1909): 167-168. MSP 104 (January 6, 1912): 85; Charles Janin, *Gold Dredging in the United States,* Bulletin No. 127 (Washington, D.C.: U.S. Bureau of Mines, 1918), p. 22; Jennings, *History of Dredging,* p. 7.

3. Thomas A. Rickard, *Interviews with Mining Engineers* (San Francisco: Mining and Scientific Press, 1922), p. 377; Allen Johnson et al., eds., *Dictionary of American Biography* [DAB], vol. 13 (New York: Scribner's, 1943), p. 135.

4. Rickard, *Interviews,* pp. 373-375.

5. Ibid., pp. 376-378; DAB, vol. 13, p. 135.

6. Not much is know about these two. According to the Boston city directories, Bigelow was part of the brokerage firm of A. M. Bigelow & Company and died in 1919. Bixby had been educated at Nashua and at Philips Andover. A fine tennis player and one-time champion rifle sharpshooter of New England, he entered business as the treasurer of the Nashua Lock Company and later held the same office in the Nashua Card and Glazed Paper Company. He died early in 1901. See *Boston Evening Transcript,* March 8, 1901.

7. See *Alder Gulch Times,* March 2, 1900; Michael Leeson, ed., *History of Montana, 1739-1885* (Chicago: Warner, Beirs & Company, 1885), p. 783; Society of Montana Pioneers, *Constitution, Members and Officers with Portraits and Maps,* vol. 1 (Akron, Ohio: J. U. Sanders), p. 177.

8. Rickard, *Interviews,* p. 378; Hennen Jennings, "Report on Properties of the Conrey Placer Mining Company and Poor Farm Placer Mining Company at Ruby, Montana," vol. 14, Conrey Papers.

9. Jennings, *Report*; MSP 100 (March 19, 1910): 431.

10. Shaler's mentor, Louis Agassiz at Harvard, had made a fortune from timber and Lake Superior copper, while Charles Chandler of Columbia and William P. Blake and Benjamin Silliman, Jr., of Yale were equally active, if somewhat less prosperous, examples.

11. Nathaniel Southgate Shaler, *The Autobiography of Nathaniel Southgate Shaler* (Boston: Houghton Mifflin, 1909), p. 393. An estimated seven thousand students passed through his popular courses. In 1904, after a faculty committee decreed that Geology 4 could no longer count toward the B.A. degree, enrollment dropped from 489 to 85. See Walter Louis Berg, "Nathaniel Southgate Shaler: A Critical Study of an Earth Scientist" (Ph.D. diss., University of Washington, Seattle, 1957), pp. 164-167.

12. Shaler, *Autobiography,* pp. 396-397, 412; Berg, "Nathaniel Southgate Shaler," pp. 60, 69, 79, 90, 98, 193, 202; Hennen Jennings, Dinner Address, March 1909 (typescript, p. 3), Jennings MSS. Born in Kentucky of a prosperous family, Shaler (1841-1906) graduated from Harvard in 1862 and served briefly in the Union Army as a captain in the Fifth Kentucky Battery before ill health compelled him to retire. After two years as assistant to Agassiz and two more in study and exploration abroad, he joined the staff at Harvard in paleontology and geology. From 1873 to 1880, when he resigned

because of a new residency requirement, he was also State Geologist of Kentucky; and for the next two decades did intermittent work for the U.S. Coast Survey, the U.S. Geological Survey, and the Massachusetts Highway Commission. See DAB, vol. 9, pp. 17-18.

13. Ibid.; MSP 100 (January 1, 1910): 67; Rickard, *Interviews,* p. 379; *Eleventh Report of the Bureau of Agriculture, Labor and Industry of the State of Montana for the Year ending November 30, 1908* (1908[?]), p. 213; Shaler, *Autobiography,* p. 397. The standard study of Shaler devotes only half a paragraph to his connection with the Conrey Placer Mining Company. See Berg, "Nathaniel Southgate Shaler," p. 154. A recent intellectual biography makes no mention of the Conrey enterprise. See David N. Livingstone, *Nathaniel Southgate Shaler and the Culture of American Science* (Tuscaloosa and London: University of Alabama Press, 1987).

14. DAB, vol. 12, pp. 73-74; *New York Times,* October 20, 1903; *The National Cyclopaedia of American Biography,* vol. 10 (New York: James T. White & Company, 1900), p. 397.

15. Shaler, *Autobiography,* pp. 327-328; Nathaniel S. Shaler, "Gordon McKay," *Harvard Graduate Magazine* (June 1905), p. 575; *New York Times,* October 20, 1903.

16. Conrey Placer Mining Company, "Articles of Incorporation," December 30, 1897 (typescript), Secretary of State's Office, Helena, Montana.

17. Agreement of December 31, 1897, cited in Declaration of Trust (October 19, 1897) filed by Harry G. Bixby, Miscellaneous Record Book 58, p. 26, Madison County Courthouse, Virginia City, Montana; Account Journal (December 1901), pp. 12, 22, vol. 7, Conrey Papers.

18. Meeting of February 15, 1898, Directors' Records, p. 1, vol. 3, Conrey Papers; James Walter Goldthwait, *Diary of Western Trip 1902* (Burnsville, North Carolina: Ad Lib Press, 1982), p. 159. "A thorough Montanian" who had lived in Virginia City since 1864, Clark was regarded as "a safe and conservative counsel" and was retained by a number of the large corporations operating in Madison County. See *Progressive Men of the State of Montana* (Chicago: A. W. Bowen & Company, c. 1902), pp. 572-573.

19. Account Journal, p. 18, vol. 7, Conrey Papers.

20. Ibid., p. 19.

21. Meeting of March 24, 1898, Directors' Records, p. 3, vol. 3, Account Journal, entries for May 1901, Account Journal, pp. 3-5, vol. 7, Conrey Papers.

22. Meeting of August 2, 1898, Directors' Records, p. 7, vol. 3, Conrey Papers; "Increase of Capital Stock of Conrey Placer Mining Company," filed in Madison County, January 24, 1899, and filed with the Secretary of State, January 28, 1899, Secretary of State's Office, Helena, Montana.

23. There were 1,769 Conrey shares outstanding at $100 par value. McKay held 800 shares; Bigelow, 327; Shaler, 231; and Bixby or his executors, 411. Philip and Eva P. Moore of St. Louis owned the original issue of capital stock. See Meetings of July 24, October 23, 1900, Directors' Records (July 24, October 23, 1900), pp. 13, 19, 22, vol. 3, Conrey Papers; deed of September 29, 1900, from German Bar Mining Company to Conrey Placer Mining Company, Deed Record Book LXV, pp. 207-210, Madison County Courthouse; Certificate of Increase of Capital Stock of Conrey Placer Mining Company, filed December 24, 1900, Secretary of State's Office, Helena.

24. Meeting of March 17, 1902, Directors' Records, p. 42, vol. 3, Conrey Papers; Conrey Placer Mining Company statement, Bonds and Notes to be converted into Preferred Shares, September 30, 1902, vol. 40, Conrey Papers. See also Certificate of Increase of Capital Stock of the Conrey Placer Mining Company, filed September 22, 1905, Secretary of State's Office, Helena.

25. Deed of November 29, 1897, from Philip and Sarah V. Conrey to H. G. Bixby, guardian's deed of December 1, 1899, from Philip Conrey to H. G. Bixby, both in Deed Record Book LXII, pp. 373-374, 406-409; deed of December 3, 1897, from Oscar A. Sedman to H. G. Bixby, deed of December 16, 1897, from W. W. and Mrs. S. B. McGregory to H. G. Bixby, deed of December 23, 1897, from Oscar A. Sedman (for John E. Sedman) to H. G. Bixby, deeds (2) of December 31, 1897, from Henry and Mary Elling to H. G. Bixby, all in Deed Record Book LXV, pp. 258, 589-596, 597-603, 603-611, Madison County Courthouse.

26. Deed of June 29, 1898, from David R. and Mary A. Beck to H. G. Bixby, deed of August 8, 1899, from William Hinch and George T. Harrison to H. G. Bixby, both in Deed Record Book LXII, pp. 211, 292-293; deed of July 1, 1898, from Fabian and Victoria Legris to H. G. Bixby, Deed Record Book LX, pp. 317-318, Madison County Courthouse.

27. Quit Claim deed of July 27, 1898, from H. G. Bixby to German Bar Mining Corporation, Quit Claim deed of July 27, 1898, from H. G. Bixby to Conrey Placer Mining Company, both in Deed Record Book LX, pp. 353-356, 357-359; deed of September 13, 1900, from H. G. Bixby to Conrey Placer Mining Company, in Deed Record Book LXV, pp. 168-169, Madison County Courthouse.

28. Deed of August 8, 1900, from Augustus Sweet to Conrey Placer Mining Company, in Deed Record Book LXV, p. 101; deed of September 7, 1901, from Joseph H. and Elizabeth Carter to Conrey Placer Mining Company, Deed Record Book LXX, pp. 37-39, Madison County Courthouse.

29. Confirmation of Sale Notice of June 25, 1900, deed of June 25, 1900, from Alden J. Bennett to Gordon McKay, both in Deed Record Book LXV, pp, 1-4, 23-26; deed of February 25, 1901, from David R. and Mary A. Beck and Oscar A. Sedman to Conrey Placer Mining Company, Deed Record Book LXV, p. 368; deed of July 3, 1902, from A. N. and Sarah E. Rodgers to Conrey Placer Mining Company, Deed Record Book LXXII, p. 78; deed of October 16, 1902, from George L. and Clara M. Ramsey to Conrey Placer Mining Company, Deed Record Book LXXV, p. 265, Madison County Courthouse.

30. Account Journal (December 1901), p. 12, vol. 7, Conrey Papers; deed of December 21, 1903, from Eliza J. Van Brocklin to Conrey Placer Mining Company, Deed Record Book LXXIII, pp. 139-140; deed of May 10, 1910, from Eliza J. Van Brocklin to Poor Farm Placer Mining Company, Deed Record Book LXXX, p. 101, Madison County Courthouse.

31. Deed of October 12, 1901, from Patrick Carney, Marcus Elser, and Osman B. Varney (county commissioners) to Conrey Placer Mining Company, Deed Record Book LXX, p. 202, Madison County Courthouse; Account Journal (December 1901), p. 12, vol. 7, Conrey Papers; Declaration of Trust of October 19, 1898, by H. G. Bixby, Miscellaneous Record Book LVIII,

pp. 260-261; Declaration of Trust of November 20, 1900, by H. G. Bixby, Miscellaneous Record Book LXVII, pp. 11-12; deed of February 1906 from N. S. Shaler, trustee, to Poor Farm Placer Mining Company, Deed Record Book LXXV, pp. 212-213, Madison County Courthouse; Poor Farm Placer Mining Company, Articles of Incorporation (Recorded at the Madison County Courthouse, October 12, 1904), in Secretary of State's Office, Helena.

32. Deed of August 15, 1903, from Philip N. and Eva Moore and Winchester and Anna W. Dickerson to Conrey Placer Mining Company, deed of August 5, 1902, from Winchester Dickerson to Conrey Placer Mining Company, both in Deed Record Book LXXIII, pp. 12, 13; deed (no date, notarized September 19, 1905) from Nathaniel and S. P. Shaler, Henry M. and Jeannette L. Bigelow, and Philip N. and Eva P. Moore to Conrey Placer Mining Company, deed of October 20, 1906, from Chong On, Wong Hing, Chong Kee & Quong Lee Company to Conrey Placer Mining Company, both in Deed Record Book LXXV, pp. 39, 385; U.S. Patent Notice of October 31, 1907 (Agnes Placer Mining Claim), U.S. Patent Notice of March 1, 1910 (Pactolus Placer Mining Claim) U.S. Patent Notice of March 3, 1913 (Dorothy Placer Mining Claim), all in Deed Record Book LXXVII, pp. 77, 139-140, Madison County Courthouse; *Alder Gulch Times*, August 2, 1907.

33. Deed of August 22, 1905, from Mary B. Elling, Mabel M. and Thomas G. Hutt, Helen K. and L. M. Arey, Rita and P. H. Gohn, Carlotta and R. H. Fenner, Horace B. and Alfie Elling and Karl Elling to Conrey Placer Mining Company, Deed Record Book LXXV, p. 166, Madison County Courthouse.

34. Trial balance of February 7, 1898, German Bar Mining Company statements, vol. 36, Conrey Papers; undated, unsigned memorandum, vol. 36, Conrey Papers; EMJ 73 (May 15, 1897): 491.

35. EMJ 94 (August 3, 1912): 229-230; deed of March 5, 1913, from James H. and Mary Vanderbeck to Conrey Placer Mining Company, Deed Record Book LXXVIII, p. 20, Madison County Courthouse; Hennen Jennings, Dinner Address, March 1909 (typescript, p. 4), Jennings MSS; MSP 75 (July 31, 1897): 103.

Chapter Three: An Era of Experimentation

1. Rickard, *Interviews*, p. 379; *Eleventh Report of the Bureau of Agriculture, Labor and Industry* (1908), p. 213.

2. EMJ 63 (May 15, 1897): 491; 91 (April 22, 1911): 812.

3. Lidgerwood advertisement in EMJ 65 (June 18, 1898): 13 of advertising section. See also MSP 71 (December 21, 1895): 401; 75 (October 30, 1897): 405; Jennings, *History of Dredging*, p. 54; *Engineering* 63 (February 26, May 28, 1897): 272-273, 716.

4. See EMJ 63 (May 15, 1897): 491; 64 (October 2, 1897): 407; MSP 74 (June 5, 1897): 475; *Spokane Review*, cited in MSP 77 (August 6, 1898): 134; *Alder Gulch Times*, August 11, 1900; Eugene B. Wilson, *Hydraulic and Placer Mining* (New York: John Wiley & Sons, 1918), pp. 258-263.

5. *Alder Gulch Times*, August 4, 1899, August 11, 1900; German Bar Mining Company Records, vol. 36, Conrey Papers; EMJ 83 (June 29, 1907): 1223; Wilson, *Hydraulic and Placer Mining*, pp. 260-262.

6. See *Alder Gulch Times,* October 20, 1899, May 25, July 21, 1900; EMJ 67 (March 4, 1899): 274.

7. Rickard, *Interviews,* p. 379.

8. EMJ 83 (June 29, 1907): 1223; Aubury, *Gold Dredging in California,* p. 157; Output of May and June 1898, German Bar Mining Company Records, vol. 36, Conrey Papers.

9. Output at German Bar, German Bar Mining Company Records, vol. 36, Conrey Papers.

10. Cost record for April 1900, German Bar Mining Company Records, vol. 36, Conrey Papers; Charles Kammerer to Charles Janin, September 13, 1915), Box 3, Janin Papers.

11. Quoted in MSP 77 (November 19, 1898): 510.

12. MSP 79 (July 22, 1899): 97.

13. EMJ 64 (November 20, 1897): 607-608.

14. *Eleventh Report of the Bureau of Agriculture, Labor and Industry* (1908), p. 213. Frank F. Stanley of Boston and Swampscott was a trustee of Stanley Manufacturing, located at 117 Lincoln in Boston. See *The Boston Directory Containing the City Record, Directory of the Citizens, Business Directory and Street Directory, with Map* (Boston: Sampson, Murdock & Company, 1899), p. 1487.

15. EMJ 83 (June 29, 1907): 1223.

16. Jennings, *History of Dredging,* pp. 7-8; Trial balance of September 30, 1901, vol. 33, Conrey Papers.

17. MSP 103 (September 2, 1911): 1 of ads.; Janin, *Gold Dredging in the United States,* p. 5; *Digging Machinery* (New York: Hayward Company, 1907), p. 4. For the use of this last item, I am indebted to Professor John Thompson, Department of Geography, University of Illinois, Urbana.

18. *Digging Machinery,* pp. 5, 13-14; MSP 90 (April 22, 1905): 247; *Engineering* 63 (June 25, 1897): 847, 849.

19. Ibid., p. 98; EMJ 67 (February 11, 1899): 175; *Eleventh Report of the Bureau of Agriculture, Labor and Industry* (1908), pp. 17-18; Trial Balance of September 30, 1901, vol. 33, Conrey Papers; Jennings, *History of Dredging,* p. 9, EMJ 65 (June 4, 1898): 682; 83 (June 29, 1907): 1226; Aubury, *Gold Dredging in California,* p. 257; Charles Kammerer to Charles Janin, September 13, 1915, Box 3, Janin Papers. Engineer Hennen Jennings, who was later closely associated with the Conrey Company, reported that the company tried two of the "grab type" steam shovels. See Hennen Jennings, "Report of Properties of the Conrey Placer Mining Company and Poor Farm Mining Company at Ruby, Montana 1906" (handwritten, reproduced by Daking Publishing Company of San Francisco, n.d.), p. 17, vol. 14, Conrey Papers. Subsequently, the *Engineering and Mining World* and the Montana Bureau of Agriculture, Labor and Industry spoke of two "orange peel" excavators used by Conrey. EMJ 91 (April 22, 1911): 812; *Eleventh Report of the Bureau of Agriculture, Labor and Industry* (1908), p. 213. Other sources refer to only one Hayward machine.

20. *Alder Gulch Times,* September 15, 1899; Jennings, "Report" (1906), vol. 14, Conrey Papers; EMJ 83 (June 29, 1907): 1225-1226; Aubury, *Gold Dredging in California,* p. 257.

21. MSP 71 (December 21, 1895): 401. By the mid-1890s the Bucyrus Company of South Milwaukee had already built forty harbor and canal dredges.

See Harold F. Williamson and Kenneth H. Myers, *Designed for Digging: The First 75 Years of the Bucyrus-Erie Company* (Evanston, Illinois: Northwestern University Press, 1955), p. 41.

22. *Alder Gulch Times,* August 4, 1899.

23. Van Jon, "There's Gold in That Gravel," *Excavating Engineer* (December 1938): 608; MSP 81 (November 17, 1900): 528. Lyden was in error when he reported that the *Maggie* was originally built at Bannack "prior to 1895." See Charles J. Lyden, *The Gold Placers of Montana,* Memoir no. 26 (Butte: Montana Bureau of Mines and Geology, 1948), p. 82.

24. Archie Gibson was one of the first dredge men at Bannack and at Horse Prairie. *Dillon Examiner,* May 26, June 16, 1897, March 1, 1899.

25. *Alder Gulch Times,* August 4, September 15, 1899; MSP 78 (February 18, 1899): 185; 81 (November 17, 1900): 528; EMJ 67 (May 27, 1899): 630; W. S. Russell, "Gold Dredging," *Mining and Metallurgy* 21 (December 1900): 196.

26. Jennings, "Report" (1906), pp. 18-19, Conrey Papers; EMJ 83 (June 29, 1907): 1226; Jennings, *History of Dredging,* pp. 10, 12; Charles Kammerer to Charles Janin, September 15, 1915, Box 3, Janin Papers.

27. EMJ 67 (March 4, 1899): 276; MSP 80 (February 10, 1900): 155; Jennings, "Report" (1906), p. 18, Conrey Papers.

28. *Dillon Examiner,* March 25, 1903.

29. Jennings, *History of Dredging,* p. 10.

Chapter Four: An Interlude of Steam

1. H. G. Bixby to Philip N. Moore, January 11, 1899 (copy), vol. 40, Conrey Papers; W. S. Russell, "Gold Dredging," *Mining and Metallurgy* 21 (December 1900): 196.

2. H. G. Bixby to Gordon McKay, January 25, 1899 (copy), vol. 40, Conrey Papers.

3. H. G. Bixby to Gordon McKay, February 1, 1899 (copy), vol. 40, Conrey Papers. Of the thirteen Risdons built before 1899, nine were for work in California, three for Colorado, and one for Idaho. See handwritten list of Risdon dredges, notebook, in George P. Hurst MSS, Box 4, Stanford University Libraries, Stanford, California.

4. See Account Journal to May 1917, vol. 7, p. 23, Conrey Papers.

5. Expense Accounts, vol. 36, Conrey Papers.

6. Data from Jennings, *History of Dredging,* pp. 11, 12.

7. Ibid., pp. 13, 14; MSP 104 (January 6, 1912): 85; EMJ 83 (June 29, 1907): 1226; Charles Kammerer to Charles Janin, October 26, 1915, Box 3, Janin Papers.

8. *Eighth Report of the Bureau of Agriculture, Labor and Industry* (1902), p. 438.

9. H. E. Bixby to Philip N. Moore, January 11, 1899 (copy), vol. 40, Record of Delays, April 1900, vol. 36, Conrey Papers.

10. *Eleventh Report of the Bureau of Agriculture, Labor and Industry* (1908), p. 214.

11. EMJ 72 (August 3, 1901): 148; Jennings, "Report" (1906), p. 18, Account Journal to May 1917, pp. 3, 33, vol. 7, Accounts (1898-1901), vol. 33, Conrey Papers.

12. Jennings, "Report" (1906), p. 30, Conrey Papers.

13. EMJ 72 (August 3, 1901): 148; 83 (June 29, 1907): 1226.

14. *Eighth Report of the Bureau of Agriculture, Labor and Industry* (1902), pp. 438-439, 1226.

15. Account Journal, p. 69, vol. 7, Conrey Papers.

16. *Alder Gulch Times,* October 27, 1900; EMJ 70 (November 17, 1900): 506.

17. Account Journal to May 1917, pp. 1, 12, vol. 7, Accounts (1898-1901), vol. 33; Jennings, "Report" (1906), p. 30, Conrey Papers. Frank F. Stanley of Boston and Swampscott was a trustee of the Stanley Manufacturing Company, located at 117 Lincoln in Boston. Born in Lynn, Massachusetts, he was educated in the public schools and by 1882 was head of the Stanley Company. A member of the Boston Chamber of Commers, the Boston Historical Society, and a number of country and yacht clubs, Stanley died in 1921. See *Boston Evening Transcript,* July 19, 1921; *The Boston Directory Containing the City Record, Directory of the Citizens, Business Directory and Street Directory, with Map* (Boston: Sampson, Murdock & Company, 1899), p. 1487; Hennen Jennings, *The McKay Endowment and Applied Science at Harvard* (Washington, D.C.: privately printed, 1918), p. 4.

18. Records of meetings of July 29, November 1, December 7, 1901, Director's Records (1898-1913), pp. 32, 40, 41, vol. 3, Conrey Papers.

19. Jennings, "Report" (1906), pp. 29-30, Conrey Papers.

20. *Fourteenth Annual Report of the Inspector of Mines of the State of Montana* (Helena: Independent Publishing Company, 1902), p. 52; Roberta Carkeek Cheney, *Names on the Face of Montana* (Missoula: Mountain Press, Company, 1984), p. 3; Muriel S. Wolle, *Montana Pay Dirt: A Guide to the Mining Camps of the Treasure State* (Chicago: Sage Books, 1963), p. 41.

21. Account Journal to May 1917, p. 33, vol. 7, Conrey Papers.

22. *Alder Gulch Times,* August 2, 1902.

23. *Alder Gulch Times,* December 27, 1902.

24. *Eleventh Report of the Bureau of Agriculture, Labor and Industry* (1908), p. 214.

25. EMJ 91 (April 22, 1911): 812.

26. Jennings, *History of Dredging,* pp. 11-14.

27. Account Journal, pp. 97, vol. 7, Conrey Papers, VII; *Eleventh Report of the Bureau of Agriculture, Labor and Industies* (1908), p. 214.

28. *Alder Gulch Times,* June 16, 1899, January 5, 1906.

29. *Eighth Report of the Bureau of Agriculture, Labor and Industry* (1902), pp. 432, 433.

30. *Alder Gulch Times,* January 29, March 25, 1904; John B. Read, ed., *Montana: The Treasure State* (St. Louis: Montana World's Fair Commission, 1904), p. 60.

31. MSP 81 (July 14, 1900): 42.

32. *Alder Gulch Times,* May 2, 1903, February 19, 1904, February 10, 1905; *Mining Science* 59 (March 4, 1909): 166.

33. *Eleventh Report of the Bureau of Agriculture, Labor and Industry* (1908), p. 214; Jennings, *History of Dredging,* p. 16.

34. EMJ 91 (April 22, 1911): 812; Jennings, *History of Dredging,* p. 16.

35. Jennings, *History of Dredging,* pp. 15, 16.

36. *Alder Gulch Times,* February 26, 1904.

37. MSP 84 (September 3, 1904): 165.

38. *Mining Reporter* 52 (December 7, 1905): 576; *Eleventh Report of the Bureau of Agriculture, Labor and Industry* (1908), p. 214; *The Mining World* (Chicago), 23 (November 18, 1905): 567.

39. Meetings of October 11, 1905, August 7, 1906, Directors' Records (1898-1913): 93, 111, vol. 3, Conrey Papers; Jennings, *History of Dredging,* p. 214; *Alder Gulch Times,* October 26, 1906.

40. *Great Falls Tribune,* June 13, 1971.

41. *New York Times,* October 20, 1903; John Hays Gardiner, *Harvard* (New York: Oxford University Press, 1914), p. 212. The McKay bequest was to be used to promote applied science by supporting professorships, workshops, and laboratories, with special attention given to mechanical engineering and instruction to students with no support outside of free public education. But the full endowment would not come immediately. There were other beneficiaries and payments to Harvard would increase as they were met. Shaler's daughter Gabrielle, for example, would receive $3,000 a year for the rest of her life. See *Will and Codicils of Gordon McKay* (undated), Jennings MSS.

42. Meetings of November 17, 1903, September 12, 1904, Directors' Records pp. 71, 76, vol. 3, Conrey Papers.

43. MSP 81 (March 16, 1901): 131; DAB, vol. 17, p. 18; meetings of March 3, May 28, 1906, Directors' Records, pp. 103, 105-106, vol. 3, Conrey Papers; *Boston Herald,* April 14, 1915; Jennings, *The McKay Endowment and Applied Science at Harvard,* p. 4; Berg, "Nathaniel Southgate Shaler," pp. 156-161; *Harvard Alumni Bulletin* 52 (November 5, 1949): 155-156.

44. Rickard, *Interviews,* p. 379; meetings of March 3, May 28, 1906, Director's Records (1898-1913), pp. 103, 105-106, vol. 3, Conrey Papers.

45. Jennings, *McKay Endowment,* p. 3; Rickard, *Interviews,* pp. 223-253; DAB, vol. 10, p. 56; *San Jose Mercury,* April 20, 1887; MSP 120 (March 13, 1920): 396; EMJ 109 (March 13, 1920): 661; *Mining and Metallurgy,* no. 161 (May 1930): 1920; Alpheus F. Williams, *Some Dreams Come True* (Capetown, South Africa: Timmons, 1948): 549, 551.

46. Jennings, "Report" (1906), pp. 86-87, 94, Conrey Papers.

47. To be fair, Jennings did note that while Conrey operating costs over eight years averaged 16.5 cents a cubic yard, they had dropped to 13.15 cents a yard for the previous two years and to 11.75 cents for the first half of 1906. Jennings, "Report" (1906), pp. 81, 83-84, 94, Conrey Papers. Records indicate that although the "Report" bears Jennings's name, much of the work was done by Kenworthy. See Meeting of September 20, 1906, Directors' Records (1898-1913), p. 118, vol. 3, Conrey Papers.

48. Meetings of September 20, September 28, 1906, Directors' Records (1898-1913), pp. 118-120, 127-128, vol. 3, Conrey Papers.

49. Meeting of November 16, 1906, Directors' Records (1898-1913), pp. 134-135, vol. 3, Conrey Papers.

50. Jennings, *History of Dredging,* p. 15.

51. A. H. Wiseman to Hennen Jennings, November 12, 1914, Jennings MSS.

52. Meeting of December 13, 1906, Directors' Records (1898-1913), pp. 139-141, vol. 3, Conrey Papers.

53. Meetings of February 23, March 26, 1907, Directors' Records (1898-1913), pp. 147, 153-154, vol. 3, Conrey Papers.

54. *Alder Gulch Times,* September 14, 1906.

55. George L. Holmes, "Preliminary Report of Property and Equipment of the Conrey Placer Mining Co. at Ruby, Montana" (September 1906) in

Jennings, "Report" (1906), 2v, 3v, 5v, 8v, 9v, 12v, 14v, Conrey Papers; George L. Holmes, "Notes on Conrey Placer, Ruby, Montana (copy), Box 7, Fred Ludwig Morris MSS, Bancroft Library, University of California, Berkeley.

56. *Holmes*, "Preliminary Reports," pp. 17v, 19v-21v, 25v-28v.

57. Ibid., pp. 49v, 55v, 56v-57v, 58v.

58. Jennings, "Report" (1906), pp. 97, 101-102, Conrey Papers.

59. *Alder Gulch Times,* December 21, 1906, February 8, 1907.

60. EMJ 84 (July 13, 1907): 72.

Chapter Five: The Triumph of Electricity

1. Poor Farm Placer Mining Company, Articles of Incorporation (typescript), Recorded in Madison County Courthouse, Virginia City, October 12, 1904, copy in Secretary of State's Office, Helena.

2. Declaration of Trust, October 19, 1898, filed by Harry G. Bixby, Miscellaneous Record Book LVIII, p. 261; Declaration of Trust, November 20, 1900, filed by Harry G. Bixby, Miscellaneous Record Book LXVII, pp. 11-12; deed of October 12, 1901, from Patrick Carney, Marcus Elser, and Osman B. Varney (county commissioners) to Conrey Placer Mining Company, Deed Record Book LXX, p. 202, Madison County Courthouse; Edmunds to Stanley, November 24, 1917 (copy), Jennings MSS.

3. Deed of February 27, 1906, from Nathaniel S. Shaler to Poor Farm Placer Mining Company, Deed Record Book LXXV, pp. 212- 213, Madison County Courthouse.

4. Records of meetings of November 19, 1904, March 31, 1905, Directors' Records (1898-1913), pp. 80, 84-85, vol. 3, Conrey Papers.

5. Certificate of Increase of Capital Stock of the Conrey Placer Mining Company, recorded September 22, 1905, Secretary of State's Office, Helena; Jennings, "Report" (1906), pp. 9-10.

6. Directors' Record, November 17, 1903, p. 70, vol. 3, Conrey Papers.

7. *Alder Gulch Times,* December 16, December 23, 1904.

8. Directors' Record, September 12, 1904, p. 76, vol. 3, Conrey Papers.

9. Directors' Record, January 17, 1905, p. 82, vol. 3, Conrey Papers.

10. Sales Order No. 769, March 2, 1905, Sales Order Drawers, Marion Steam Shovel Division, Dresser Industries, Marion, Ohio.

11. *Alder Gulch Times,* May 12, 1905. See also Wolle, *Montana Pay Dirt,* p. 41. Wolle errs in saying that McKay sent Baier to California to study placer methods and that while he was gone the third dredge was built and put into operation. She is also mistaken in insisting that Baier returned to supervise construction in 1910 of the big No. 4 dredge.

12. *Alder Gulch Times,* January 19, 1906.

13. *Eleventh Report of the Bureau of Agriculture, Labor and Industry* (1908), p. 214.

14. EMJ 84 (July 13, 1907): 72.

15. Jennings, "Report" (1906), p. 37. In the autumn of 1913, after the dredge had been overhauled and improved, Jennings noted that the original cost had been about $175,000. Hennen Jennings to William W. Mein, September 11, 1913 (copy), Jennings MSS.

16. Jennings, "Report" (1906), Exhibit "V," pp. 31v-34v, 36v-37v, 39v-40v, 43v-44v.

17. MSP 92 (February 10, 1906): 81; EMJ 91 (April 11, 1911): 812.

18. MSP 100 (January 1, 1910): 67-68; *Eleventh Report of the Bureau of Agriculture, Labor and Industry* (1908), p. 215; Aubury, *Gold Dredging in California*, pp. 259, 262; EMJ 84 (July 13, 1907): 70.

19. EMJ 84 (July 13, 1907): 71, 72.

20. Ibid., p. 70; Jennings, *History of Dredging*, p. 21.

21. *Alder Gulch Times*, February 1, 1908; Jennings to Edward Benjamin, March 5, 1908 (copy), Jennings MSS.

22. MSP 100 (January 1, 1910): 68; EMJ 91 (April 11, 1911): 812.

23. Frank F. Stanley to Johnson & Higgins, July 26, 1915 (copy), vol. 16, Conrey Papers; MSP 103 (November 25, 1911): 676; 104 (January 6, 1912): 85.

24. *Alder Gulch Times*, January 14, February 11, 1910, January 12, 1912, March 31, 1916.

25. Janin, *Gold Dredging in the United States*, p. 153; Charles Kammerer Report, July 2, 1909, in Report of Mr. Hennen Jennings Conrey Placer Mining Co. and Poor Farm Placer Mining Co., July 1909 (typescript copy), p. 3-K, vol. 41, Conrey Papers; Frank F. Stanley to Philip N. Moore, July 5, 1915 (copy), vol. 16, Conrey Papers. Dredging was more easily handled while digging upstream because tailings could be used more readily to form effective dams to achieve flotation.

26. Stanley to Charles Kammerer, July 15, 1915, Stanley to Johnson & Higgins, August 11, 1915 (copy), vol. 16, Conrey Papers.

27. Charles Kammerer to Frank F. Stanley, November 23, 1916, vol. 27, Conrey Papers. For efforts to sell the machinery from No. 3, see G. H. Edmunds to Stanley, December 31, 1918, vol. 29, George J. Carr to Hennen Jennings, May 26, 1919, vol. 30, Record of meeting of June 23, 1919, Director's Records, pp. 171-172, vol. 4, Conrey Papers.

28. *Eleventh Report of the Bureau of Agriculture, Labor and Industy* (1908), p. 215; George L. Holmes, "Notes on Conrey Placer, Ruby, Montana," n.p., n.d. (copy), Fred L. Morris MSS, Box 7, Bancroft Library.

29. Orders No. 659 & 660, June 29, 1907, in Sales Order Drawers, Marion Power Shovel Division, Dresser Industries; Jennings to William W. Mein, September 11, 1913 (copy), Jennings MSS; *Mining Science* 59 (March 4, 1909): 167.

30. Ibid.; *Alder Gulch Times*, March 29, 1907; *Eleventh Report of the Bureau of Agriculture, Labor and Industry* (1908), p. 215.

31. Meeting of February 27, 1908, Director's Records, pp. 181-182, vol. 3, Conrey Papers.

32. Machine Record Book No. 2, p. 203, Marion Power Shovel Division; Sales Department Repair Order No. 1124, March 3, 1908, with Order No. 659, Waddell Memo, March 7, 1908; Sales Department Repair Order No. 905, February 20, 1908, with Order No. 660, June 29, 1907, all in Sales Order Drawers, Marion Power Shovel Division; *Alder Gulch Times*, October 4, 1907. How many local men were employed is not noted. Along with machinery itself, the Marion Company sent a large supply of materials to be used on the hulls—wire spikes, oakum, linseed oil, and paint. See Sales Department Repair Order No. 6238, August 10, 1907, with Order No. 660, Marion Power Shovel Division.

33. *Eleventh Report of the Bureau of Agriculture, Labor and Industry* (1908), p. 217.

34. Order No. 660, June 29, 1907, Sales Order Drawers, Marion Power Shovel Division.

35. Jennings, *History of Dredging,* pp. 11-12; *Eleventh Report of the Bureau of Agriculture, Labor and Industry* (1908), p. 215; *Mining Science* 59 (March 4, 1909): 167; EMJ 91 (April 22, 1911): 814.

36. Sales Department Repair Order No. 1124, March 3, 1908, with Order No. 659, Sales Order Drawers, Marion Power Shovel Division. There were seven other motors, ranging from 10 to 75 horsepower; an elaborate switchboard; circuit breakers; a variety of meters for voltage, amperage, and wattage; and four transformers, three of which reduced the voltage from 2,200 to 440. See Waddell Memo, August 10, 1907, with Order No. 660, ibid.

37. *Mining Science* 59 (March 4, 1909): 167.

38. *Eleventh Report of the Bureau of Agriculture, Labor and Industry* (1908), p. 216; Jennings, *History of Dredging,* p. 11.

39. *Eleventh Report of the Bureau of Agriculture, Labor and Industry* (1908), p. 215; *Mining Science* 59 (March 4, 1909): 167; MSP 100 (January 1, 1910): 68; Orders No. 659 and No. 660, June 29, 1907, Sales Order Drawers, Marion Power Shovel Division.

40. *Mining Science* 59 (March 4, 1909): 167.

41. EMJ 91 (April 22, 1911): 814.

42. J. P. Rowe, "Gold Dredging Operations in Montana," *Mining World* 33 (August 27, 1910): 350.

43. The Yuba Construction Company of California had bid $313,000 for the job, but Marion came in considerably under that. *Alder Gulch Times,* March 4, 1910; Report of Hennen Jennings, Conrey and Poor Farm Placer Mines, July 1913 (typescript), p. 10-J, Conrey Papers; Jennings to Ross Browne, March 30, 1910, Box 3, Ross Hoffmann MSS, American Heritage Center, University of Wyoming, Laramie.

44. Order No. 1226, March 3, 1910, Sales Order Drawers, Marion Power Shovel Division; MSP 100 (March 26, 1910): 469; EMJ 89 (February 12, 1910): 391.

45. R. J. Whitcoff Memo, April 26, 1910, in Order No. 1226, Sales Order Drawers, Marion Power Shovel Division.

46. *Alder Gulch Times,* March 4, 1910; EMJ 91 (April 22, 1911): 815; George W. King, "The Marion Steam Shovel Company Family," in *History of Ohio,* ed. Emi Oviatt Randall and Daniel J. Ryan, vol. 5 (New York: Century History Company, 1915), p. 80.

47. Specification of a 15-cu-ft. Elevator Dredge No. 4 for Poor Farm Placer Mining Company, February 18, 1910 (copy), Book of Specifications, Miscellaneous Files, Marion Steam Power Division. Hennen Jennings wrote that the contract price for No. 4 was "something like 230,000" but that by the fall of 1913 it had cost over $270,000. Officially, Jennings reported that $16,746.75 had to be added to the contract price of $230,000 for starting up, plus interest during construction, which together brought the total to $262,643.75. See Jennings to William W. Mein, September 11, 1913 (copy), Jennings MSS; Report of Hennen Jennings, Conrey and Poor Farm Placer Mines, July, 1913 (typescript), p. 10-J, vol. 41, Conrey Papers. Wolle, *Montana Pay Dirt,* p. 41, is far off the mark in setting the cost at "close to half a million dollars."

48. *Alder Gulch Times,* April 8, 1910.

49. *Alder Gulch Times,* July 8, October 28, 1910.

50. *Alder Gulch Times,* January 13, January 20, 1911. Another Marion employee, J. D. Shelton, apparently brought his wife with him for several months.

51. King, "The Marion Steam Shovel Company Family," pp. 79-80.

52. Hennen Jennings to Sidney Jennings, April 26, 1911 (copy), Jennings MSS; *Alder Gulch Times,* April 14, 1911.

53. *Alder Gulch Times,* April 28, 1911; Hennen Jennings to William W. Mein, May 12, 1911 (copy), Box 6, Jennings MSS.

54. Specification of a 15 cu-ft. Elevator Dredge No. 4 for Poor Farm Placer Mining Company, February 18, 1910 (copy), Book of Specifications, Miscellaneous Files, Marion Steam Power Division; EMJ 91 (April 22, 1911): 815.

55. Ibid.

56. Ibid.; Jennings, *History of Dredging,* p. 14.

57. Specification of a 15 cu-ft. Elevator Dredge No. 4 for Poor Farm Placer Mining Company, February 18, 1910 (copy), Book of Specifications, Miscellaneous Files, Marion Steam Power Division; EMJ 91 (April 22, 1911): 815.

58. Ibid., MSP 100 (March 26, 1910): 469.

59. Ibid.

60. MSP 103 (November 25, 1911): 676; EMJ 91 (April 22, 1911): 815.

61. EMJ 90 (April 2, December 24, 1910): 741, 1277.

62. EMJ 95 (June 7, 1913): 1171; MSP 106 (January 4, 1913): 44; 108 (January 10, 1914): 96; Hennen Jennings to G. H. Edmunds, January 21, 1918 (copy), Jennings to William W. Mein, September 11, 1913 (copy), both in Jennings MSS; G. H. Edmunds and Oscar O. Sweeny to Jennings, January 13, 1918, vol. 29, Conrey Papers; Jennings to Charles Janin, November 30, 1912, Box 15, Charles Kammerer to Charles Janin, September 15, 1915, Box 3, Charles Janin MSS; Report of Hennen Jennings, Conrey and Poor Farm Placer Mines, July 1913 (typescript), vol. 41, Conrey Papers. The liquid rheostat switch control was not entirely new, having had been used by the Natomas Company in California; but Conrey developed a much more sophisticated and effective apparatus. See excerpt from paper read at the American Institute of Electrical Engineers (August 1914), EMJ 98 (November 7, 1914): 827-828.

63. EMJ 98 (November 14, 1914): 876; 99 (January 9, 1915): 110; Report of Hennen Jennings, Conrey and Poor Farm Placer Mines, July 1913 (typescript), vol. 41, Conrey Papers.

Chapter Six: Navigating Alder Gulch

1. *Alder Gulch Times,* March 15, 1902; MSP 84 (April 19, 1902): p. 224; Jennings, "Report" (1906), p. 31; Record of Directors' meeting, September 16, 1902, p. 52, Record of Directors' meeting, November 1, 1902, p. 54, vol. 3, Conrey Papers.

2. *Alder Gulch Times,* April 17, 1908; *Mining Science* 59 (March 4, 1909): 167.

3. Jennings to W. W. Mein, September 11, 1913 (copy), Jennings MSS.

4. Jennings, *History of Dredging,* p. 20.

5. Edmunds to Henry Bigelow, June 12, 1917, Jennings MSS.

6. EMJ 99 (June 26, 1915): 1139; F. F. Stanley to W. L. Webb, June 19, 1915 (copy), vol. 16, Conrey Papers.

7. Edmunds to F. F. Stanley, November 24, 1917 (copy), Edmunds to Jennings, February 22, 1918, Edmunds and Sweeny to Jennings, July 15, 1917, Jennings MSS.

8. *Eleventh Report of the Bureau of Agriculture, Labor and Industry* (1908), p. 217.

9. Rickard, *Interviews*, p. 380. See also *Madisonian Times*, December 3, 1915.

10. *Alder Gulch Times,* December 27, 1902.

11. *Alder Gulch Times,* February 21, 1902.

12. EMJ 77 (May 26, 1904): 846.

13. George L. Holmes, "Notes on Conrey Placer, Ruby, Montana," n.d. (copy), Box 7, Fred Ludwig Morris MSS, Bancroft Library, University of California, Berkeley; "Yield & Working Costs in Connection with Temperatures, January 1905-May 1906" (typescript, copy), Box 3, Charles Janin MSS; Jennings, "Report" (1906), p. 21; Aubury, *Gold Dredging in California,* pp. 261-262; EMJ 83 (January 5, 1907): 23; 84 (July 13, 1907): 71.

14. Report of Charles Kammerer, July 2, 1909, in Report of Mr. Hennen Jennings, Conrey Placer Mining Company and Poor Farm Placer Mining Company, July 1909 (typescript), p. 1-K, vol. 41, Conrey Papers.

15. Edmunds to Stanley, December 31, 1918, vol. 29, Conrey Papers.

16. *Madisonian Times,* August 27, 1915.

17. Charles Kammerer to Charles Janin, October 26, 1915, Box 3, Charles Janin MSS; Charles Kammerer Report, July 2, 1909, in Report of Mr. Hennen Jennings, Conrey Placer Mining Company and Poor Farm Placer Mining Company, July 1909, p. 2-K (typescript), vol. 41, Conrey Papers; Edmunds, Sweeny and Prier to Jennings, August 8, 1917, Jennings MSS; Record of Delays, pp. 61, 69, 70, 80, 94, 108, 178, 179, vol. 13, Conrey Papers.

18. Edmunds to Willoughby L. Webb, June 16, 1918 (copy), Edmunds to Stanley, June 15, 1918 (copy), Edmunds and Sweeny to Jennings, June 4, 1918, Jennings MSS; Edmunds to Jennings, July 17, 1918, vol. 29, Conrey Papers.

19. Record of Delays, pp. 1, 9, 17, 49, vol. 13, Conrey Papers.

20. Ibid., pp. 25, 33, 41, 48.

21. Jennings, "Report" (1906), pp. 86-87; Record of Delay, May, June, and July 1900, vol. 33, Conrey Papers.

22. Report of Delays, *Gordon* Cost Sheet, September 1900, vol. 3, Report of Delays, *Gordon* Cost Sheets, October and November 1900, vol. 33, Conrey Papers.

23. Record of Delays, pp. 22, 23, 29, vol. 13, Conrey Papers; *Alder Gulch Times,* April 5, 1907.

24. *Alder Gulch Times,* June 17, 1904; Edmunds and Sweeny to Jennings, January 13, 1918, vol. 29, Conrey Papers; Edmunds and Sweeny to Jennings, January 18, January 20, 1918, Jennings MSS.

25. Record of Delays, p. 55, vol. 13, Conrey Papers.

26. Report of Charles Kammerer, July 2, 1909, in Report of Mr. Hennen Jennings, Conrey Placer Mining Company and Poor Farm Placer Mining Company, July 1909, p. 1-K (typescript), vol. 41, Conrey Papers.

27. Record of Delays, pp. 22, 23, 29, vol. 13, Conrey Papers.

28. Report of Hennen Jennings, Conrey and Poor Farm Placer Mines, July 1913 (typescript), p. 8-J, vol. 41, Conrey Papers; EMJ 91 (February 4, 1911): 290; *Alder Gulch Times,* May 19, 1911; *The Salt Lake Mining Review* 15 (September 15, 1913): 16.

29. Ibid.; Report of Hennen Jennings, Conrey and Poor Farm Placer Mines, July 1913 (typescript), p. 8-J, vol. 40, Conrey Papers.

30. *Alder Gulch Times,* April 10, 1914.

31. Jennings to Stanley, November 21, 1916; Charles Kammerer to Jennings, December 5, 1916 (copy), vol. 27, Conrey Papers.

32. *Alder Gulch Times,* November 27, 1914; Statement of Delays, week ending December 18, 1915, in Letters Received File, vol. 26, Conrey Papers.

33. Edmunds to Jennings, July 20, 1917, Jennings to Edmunds, July 26, 1917 (copy), Jennings MSS.

34. Edmunds to Jennings, June 7, 1917, Jennings MSS. Jennings's advice was not to bother with the lead until the company decided what to do with the large amounts of black sand and iron it had on hand. See Jennings to Edmunds, June 23, 1917 (copy), Jennings MSS.

35. Edmunds to Willoughby L. Webb, October 12, 1917, June 16, 1918 (copies), Jennings MSS.

36. Edmunds to Stanley, June 22, 1918, vol. 29, Conrey Papers; Edmunds to Jennings, October 12, 1917, June 24, 1918; Edmunds and Sweeny to Jennings, October 12, 1917, January 18, 1918, Jennings MSS.

37. Edmunds, Sweeny, and Prier to Jennings, June 2, 1917; Jennings to Edmunds, June 23, 1917 (copy), Jennings MSS.

38. Edmunds to Johnson and Higgins, August 2, 1917 (copy), letterpress book, vol. 16, Conrey Papers; Edmunds to Stanley, July 27, 1917; Edmunds to Jennings, July 27, 1917, Jennings MSS.

39. Cost of the coal at the mine was $4.25 a ton. Freight from the mine to Alder by rail added another $2.20, and it cost at least $1.00 to haul the coal by traction engine and store it at the dredge. See Edmunds to Stanley, August 14, 1917, Jennings MSS.

40. Edmunds to Jennings, July 27, 1917, May 5, June 24, 1918, Jennings to Edmunds, July 29, 1917 (copy), Jennings MSS; Edmunds to Stanley, June 22, 1918, vol. 29, Conrey Papers.

41. *Alder Gulch Times,* October 28, 1910.

42. Edmunds to Stanley, November 8, 1918, vol. 29, Conrey Papers.

43. Earlier, W. W. Parker had bought the old *Maggie* at a sheriff's sale and was notified by Conrey officials to remove it from the property. Instead, Parker took what he wanted and left the remainder. In 1918, the company sought to assess him the cost of removing the relic. See Edmunds to Stanley, January 21, 1918, Jennings MSS.

44. Edmunds to Jennings, July 17, 1918, Edmunds and Sweeny to Stanley, August 1, 1918 (copy), Conrey Papers.

45. Edmunds and Sweeny to Jennings, January 18, 1918, Jennings MSS.

46. Edmunds and Sweeny to Jennings, January 18, January 31, 1918, Jennings MSS.

47. Edmunds and Sweeny to Jennings, January 31, 1918, Jennings MSS.

48. Edmunds and Sweeny to Jennings, January 20, February 19, 1918, Jennings MSS.

49. Edmunds and Sweeny to Jennings, May 15, 1918, Jennings MSS.

50. Jennings, *History of Dredging,* p. 29.

51. Report of T. D. Prier, July 2, 1909, in Report of Mr. Hennen Jennings, Conrey Placer Mining Company and Poor Farm Placer Mining Company, July 1909, pp. 1-P-2-P (typescript), Conrey Papers; Jennings to W. W. Mein, September 11, 1913 (copy), Jennings MSS; Janin, *Gold Dredging in the United States,* p. 158.

52. Edmunds to Stanley, May 27, 1920, Conrey Papers.

53. *Alder Gulch Times,* December 11, 1914.

54. *Madisonian Times,* August 27, 1986.

55. Edmunds to Henry M. Bigelow, June 12, 1917 (copy), Jennings MSS.

56. Edmunds and Sweeny to Jennings, January 13, 1918, vol. 29, Conrey Papers; Edmunds and Sweeny to Jennings, January 18, January 20, January 31, February 19, 1918, Jennings to Edmunds, January 21, 1918 (copy), Jennings MSS.

57. Weatherbe, *Dredging for Gold in California,* pp. 31-32; MSP 121 (October 16, 1921): 546; Janin, *Gold Dredging in the United States,* p. 28; T. C. Earl, *Gold Dredging* (London: E. & F. N. Spon, 1913), pp. 58-61.

58. A. H. Wiseman to editor, March 1, 1910, MSP 100 (March 19, 1910): 431; (November 25, 1899), p. 648; Account Journal, December 1901, p. 33, vol. 7, Conrey Papers; *Alder Gulch Times,* March 9, 1900.

59. Directors' Meeting of November 17, 1903, p. 70, vol. 3, Conrey Papers.

60. EMJ 83 (June 29, 1907): 1225.

61. Edmunds to Jennings, June 19, 1917, March 28, 1918; Edmunds and Sweeny to Jennings, May 15, 1918, Jennings MSS.

62. Edmunds, Prier, and Sweeney to Jennings, July 21, 1917, Jennings MSS.

63. See deed of September 5, 1905, from E. Munroe to Conrey Placer Mining Company, Deed Record Book LXIX, p. 426, deed of February 10, 1908, from Gertrude Thompson, Rita Perkins, and William B. Thompson to Conrey Placer Mining Company, Deed Record Book LXXIX, pp. 164-165, deed of June 26, 1911, from J. C. and Cornelia A. Fitzpatrick to Poor Farm Placer Mining Company, Deed Record Book LXXX, pp. 302, 303, deed of June 26, 1912, from Laura R. Duncan, William Fehring, and James and Elsie Garrison to Conrey Placer Mining Company, Deed Record Book LXXXVI, pp. 378-380, deed of March 5, 1913, from James H. and Mary Vanderbeck to Conrey Placer Mining Company, Deed Record Book LXXXVIII, p. 20, all in Madison County courthouse; Stanley to Charles Kammerer, October 19, 1915 (copy), letterbook, vol. 16, Conrey Papers.

64. *Alder Gulch Times,* May 3, May 10, 1907.

65. *Alder Gulch Times,* September 10, 1909.

66. *Alder Gulch Times,* October 8, 1909; EMJ 88 (October 16, 1909): 796; Jennings to Ross Hoffmann, April 8, April 29, 1910 (copies), Hoffmann to Jennings, April 21, May 5, 1910 (drafts), Kammerer to Hoffmann, May 2, 1910 (copy), Box 3, Hoffmann MSS.

67. *Alder Gulch Times,* October 21, 1910, July 24, 1914.

68. *Madisonian Times,* August 20, October 1, 1915.

69. Edmunds to Willoughby L. Webb, October 12, 1917 (copy), Edmunds to Jennings, June 16, 1918, Jennings MSS.

70. Jennings, *History of Dredging,* pp. 28, 29-30.

71. Janin, *Gold Dredging in the United States,* p. 190.

Chapter Seven: Of Men and Machines

1. Shaler visited Alder Gulch in the summer of 1897 and in 1902 and 1904. See Shaler, *Autobiography,* pp. 336, 397-398, 408, 412; Garthwait, *Diary of a Western Trip,* p. 155.

2. Edmunds and Sweeny to Stanley, January 19, 1918 (copy), Edmunds to Stanley, November 24, 1917 (copy), Stanley to Edmunds, February 8, 1918 (copy), Jennings MSS.

3. Edmunds and Sweeny to Jennings, January 31, 1918, Edmunds to Jennings, February 22, 1918, Jennings MSS.

4. *Alder Gulch Times,* October 17, 1900, May 12, 1905; Garthwait, *Diary of a Western Trip,* p. 162.

5. Meeting of June 5, 1905, Directors' Records, pp. 86-87.

6. *Alder Gulch Times,* July 5, 1902, April 14, May 26, 1905, June 26, 1908.

7. Hennen Jennings to Sidney Jennings, April 26, 1911 (copy), Jennings MSS.

8. Jennings to Charles H. McNary, May 19, 1910, Jennings MSS.

9. *Alder Gulch Times,* June 26, 1908; meeting of July 7, 1908, Directors' Records, p. 190, vol. 3, Conrey Papers.

10. A. H. Wiseman to Jennings, June 18, 1909, January 30, February 14, 1912, Jennings MSS; *Seattle City Directory for the Year Commencing May 1st 1914* (Seattle: R. L. Polk & Company, 1914), p. 1789.

11. *Alder Gulch Times,* June 26, 1908.

12. Jennings to Adams, May 10, 1917, vol. 44, Conrey Papers.

13. Jennings to Edmunds, April 12, 1917, Jennings MSS.

14. Thomas H. Leggett to Jennings, May 16, 1917, vol. 44, Conrey Papers.

15. Jennings to Edmunds, Prier and Sweeny, May 10, 1917 (copy), Jennings MSS.

16. Ibid.

17. Edmunds to Jennings, June 29, 1917; Thomas H. Leggett to Edmunds, June 25, 1917 (copy), Jennings MSS.

18. Prier's job ended in mid-October 1917, but his salary ran to the end of the year. When he left, he sold to Edmunds the three Conrey shares he held for $100 each. See Edmunds to Jennings, August 11, October 12, 1917, Edmunds to Stanley, October 11, 1917 (copy), Edmunds to Willoughby L. Webb, October 12, 1917 (copy), Jennings MSS; Directors' Records, September 21, 1917 (copy), p. 128, vol. 4, Conrey Papers; Jennings to Charles Francis Adams, November 20, 1917, Docket Filed Material, Estate of Gordon McKay, vol. 44, Conrey Papers.

19. Jennings to Edmunds, June 29, 1917 (copy), Edmunds to Jennings, July 5, 1917, March 21, April 30, 1918, Edmunds and Sweeny to Jennings, March 25, 1918 (personal), Jennings MSS; *Madisonian Times,* August 9, 1918.

20. Edmunds to Stanley, March 13, 1920, vol. 31, Conrey Papers; Edmunds to Charles Janin, April 27, September 9, 1936, Box 4, Janin MSS. When he died, Jennings left an estate of at least $2 million. See EMJ 109 (March 27, 1930): 740; *Madisonian Times,* March 12, 1920.

21. *Alder Gulch Times,* May 2, 1903.

22. MSP 103 (November 25, 1911): 677.

23. In his 1916 description of Conrey operations, Hennen Jennings wrote that steam No. 1 and No. 2 each employed fifteen men, but he gave no breakdown and there is no way of knowing how many of these were on shore or incidental workers. Jennings, *History of Dredging,* p. 20.

24. MSP 103 (November 25, 1911): 677.

25. Lela Gilman, "Alder Gulch" (Montana State University, Bozeman, 1930), p. 8. Used courtesy of Montana State University Library.

26. *Eleventh Report of the Bureau of Agriculture, Labor and Industry* (1908), p. 217.

27. Jennings, *History of Dredging,* p. 20. Apart from shoremen and auxiliary help, by 1917, after No. 3 was retired, the actual on-the-dredge crew of a given shift on electrics No. 1 and No. 2 consisted of dredgemaster, pilotman, lineman, and one oiler. On the larger No. 4, there was an additional oiler. See copy of "Schedule on Wages," with Edmunds and Sweeny to Jennings, July 15, 1917, Jennings MSS.

28. Edmunds and Sweeny to Jennings, June 4, 1918, Jennings MSS.

29. Edmunds to Jennings, June 7, 1917, Jennings MSS.

30. Edmunds and Sweeny to Jennings, July 15, 1917, Jennings MSS.

31. Edmunds and Sweeny to Jennings, July 15, 1917, Jennings MSS; *The Madisonian,* January 28, 1955; *Great Falls Tribune,* June 13, 1971; MSP 103 (November 25, 1911): 77.

32. Edmunds and Sweeny to Jennings, March 25, 1918 (personal), Jennings MSS.

33. Edmunds to Jennings, June 7, 1917, Jennings MSS.

34. *Alder Gulch Times,* December 11, 1908, February 5, February 12, March 19, March 26, September 10, September 24, 1909.

35. *Alder Gulch Times,* April 8, 1904.

36. *Alder Gulch Times,* August 17, November 2, 1906, November 12, 1909. During the wartime manpower shortage, Conrey contemplated hiring the Madison County surveyor to work two or three days at the beginning of each month to get yardage and property maps in shape, but it probably did not go ahead with that part-time arrangement. See Edmunds to Jennings, June 19, 1917, Jennings MSS.

37. EMJ 99 (February 27, 1915): 396.

38. *Alder Gulch Times,* March 18, 1904.

39. *Alder Gulch Times,* May 11, 1906; Charles Cassell to Jennings, October 10, 1907, Jennings MSS.

40. *Alder Gulch Times,* August 17, 1906, March 14, 1913.

41. *Alder Gulch Times,* August 2, 1902.

42. *Alder Gulch Times,* November 1, 1902.

43. Meeting of November 17, 1903, Directors' Records, p. 70, vol. 3, Conrey Papers.

44. *Alder Gulch Times,* July 10, 1903.

45. *Alder Gulch Times,* February 10, 1905.

46. *Alder Gulch Times,* June 9, 1905.

47. *Alder Gulch Times,* January 10, 1908.

48. *Alder Gulch Times,* May 8, 1908.

49. Meetings of October 11, 1909, March 14, 1910, Directors' Records, pp. 219-220, 233-234, vol. 3, Conrey Papers.

50. *Alder Gulch Times,* December 24, December 31, 1909.

51. Meetings of March 14, 1910, June 2, 1911, Directors' Records, pp. 233-234, 242-243, vol. 3, Conrey Papers.

52. *Alder Gulch Times,* March 3, March 24, 1911.

53. *Alder Gulch Times,* September 15, 1911.

54. *Alder Gulch Times,* August 2, 1912.

55. *Alder Gulch Times,* January 21, 1914.

56. *Alder Gulch Times,* December 11, 1914; MSP 109 (December 26, 1914): 1008.

57. Charles Kammerer to Jennings, January 28, 1915 (copy), vol. 26, Conrey Papers.

58. Kammerer to Jennings, December 21, 1914 (copy), vol. 26, Conrey Papers.

59. *Alder Gulch Times,* October 29, 1915, March 31, 1916.

60. Edmunds to F. F. Stanley, December 11, 1917, Jennings MSS.

61. Edmunds to Hennen Jennings, December 13, 1918, vol 29, Conrey Papers; *Madisonian Times,* December 6, 1918.

62. Meeting of March 14, 1910, Directors' Records, pp. 230- 231, vol. 3, meeting of May 14, 1915, Directors' Records, p. 37, vol. 4, Conrey Papers.

63. Edmunds and O. O. Sweeny to Jennings, May 15, 1918, Jennings MSS.

64. Meeting of December 7, 1901, Directors' Records, p. 41, vol. 3, Conrey Papers.

65. Jennings, "Report" (1906), pp. 32, 110-111.

66. Ibid., pp. 13-14.

67. Meeting of August 23, 1906, Directors' Records, pp. 115-116, vol. 3, Conrey Papers.

68. EMJ 84 (July 13, 1907): 72.

69. *Alder Gulch Times,* February 12, 1909.

70. *Alder Gulch Times,* December 30, 1910.

71. Stanley to R. Stuyvesant Pierrepont, October 30, 1915 (copy), letterbook, vol. 16, Conrey Papers.

72. Jennings to Charles Francis Adams II, April 29, 1910 (copy), Jennings to Charles H. McNary, May 19, 1910 (copy), Jennings to Edmunds, March 13, 1918 (copy), Jennings MSS.

73. Edmunds to Jennings, January 24, 1918, Jennings MSS.

74. Edmunds, Sweeny and Prier to Jennings, June 2, 1917, Jennings MSS.

75. Edmunds to Jennings, June 19, 1917, Edmunds and Sweeny to Jennings, July 15, 1917, Jennings MSS.

76. Edmunds and Sweeny to Jennings, January 31, 1918, Jennings MSS.

77. Edmunds and Sweeny to Jennings, May 17, 1918, Jennings MSS.

78. Edmunds and Sweeny to Jennings, May 15, 1918; Edmunds to Jennings, April 4, 1918, Jennings MSS.

79. Edmunds and Sweeny to Jennings, June 2, 1917, Jennings MSS.

80. Edmunds and Sweeny to Jennings, July 15, 1917, Jennings MSS.

81. Edmunds to Jennings, June 19, July 10, August 11, 1917, Edmunds and Sweeny to Jennings, June 4, 1918, Jennings MSS.

82. Edmunds and Sweeny to Jennings, July 15, 1917, Jennings MSS.

83. Edmunds to Jennings, June 19, June 29, July 5, 1917, Jennings MSS.

84. Edmunds to Jennings, May 18, 1917, Jennings MSS.

85. Edmunds to Jennings, December 3, 1917, Jennings MSS.

86. One traveling man said he thought "she wore ribbons on her night shirt." See Edmunds to F. F. Stanley, February 11, 1918 (personal), vol. 29, Conrey Papers.

87. Edmunds to Jennings, February 19, 1918, Jennings MSS; Edmunds to Stanley, November 8, 1918, vol. 29, Conrey Papers.

88. Edmunds and Sweeny to Jennings, June 1, 1917, Jennings to Edmunds and Sweeny, June 9, 1917, Jennings MSS.

89. Jennings to Edmunds, July 26, 1917 (copy), Jennings MSS.

90. Edmunds and Sweeny to Jennings, June 27, 1917, Jennings MSS.

91. Edmunds and Sweeny to Jennings, July 15, 1917, Jennings MSS.

92. Edmunds to Stanley, October 11, 1917, vol. 28, Conrey Papers.

93. Jennings to Stanley, December 15, 1917, vol. 28, Conrey Papers; Jennings to Edmunds, January 12, 1918 (copy), Jennings MSS.

94. Edmunds to Stanley, April 5, 1918 (copy), Jennings MSS.

95. Edmunds and Sweeny to Jennings, May 17, 1918, Jennings MSS.

96. Jennings to Edmunds and Sweeny, May 23, June 11, 1918 (copies), Jennings MSS.

97. Edmunds and Sweeny to Jennings, June 2, 1918, vol. 29, Conrey Papers.

98. Included in the raises were: dredgemasters, from $198 a month to $210; pilotmen, from $5.50 to $6.00 for eight hours; oilers, from $4.40 to $4.80 for eight hours; linemen, from $4.40 for nine hours to $4.50 for eight hours; machinists from $5.50 for nine hours to $5.60 for eight hours; electricians and clean-up crew, $165 a month on a nine-hour basis to $170 a month on an eight-hour basis; watchmen, $4.40 to $4.84 for a ten-hour shift, except for May through August, when nine hours prevailed. See Edmunds and Sweeny to Jennings, June 4, 1918, Jennings MSS.

99. Edmunds to Stanley, June 15, 1918, vol. 29, Conrey Papers.

100. Edmunds to Jennings, August 24, 1918 (copy), vol. 29, Conrey Papers.

101. Jennings to F. F. Stanley, December 30, 1918, Edmunds to Stanley, December 31, 1918, vol. 29, Directors' Records, December 23, 1919, p. 179, vol. 4, Jennings to Stanley, December 18, 1919, vol. 30, Conrey Papers.

102. Edmunds to Stanley, November 20, 1919, vol. 30, Conrey Papers.

103. Record of Meeting, September 3, 1920, p. 184, vol. 4, Conrey Papers.

104. Edmunds to Stanley, September 14, 1920, vol. 31, December 31, 1920, pp. 184, 195, vol. 4, Conrey Papers.

105. Charles Francis Adams II to Stanley, May 20, 1921, vol. 32, Conrey Papers.

106. Cheney, *Names on the Face of Montana*, p. 233; Meeting of September 3, 1902, Directors' Records, p. 50, vol. 3, Conrey Papers.

107. Meeting of November 8, 1905, Directors' Records, p. 95, vol. 3, Conrey Papers.

108. *Alder Gulch Times*, January 19, 1906; *Mining Science* 59 (March 4, 1909): 167.

109. Wolle, *Montana Pay Dirt*, p. 41; *Great Falls Tribune*, June 13, 1971.

110. Manuscript, Census of 1910, Madison County, Montana.

111. Edmunds to Jennings, August 11, 1917, Jennings MSS.

112. Edmunds to Stanley, November 24, 1917 (copy), Jennings MSS. See also insurance statement accompanying Edmunds to Jennings, February 22, 1918, Jennings MSS.

113. *Alder Gulch Times*, November 20, 1903.

114. Edmunds and Sweeny to Jennings, June 27, 1917, Jennings MSS.

115. Edmunds and Sweeny to Jennings, May 15, 1918, Jennings MSS.

116. Directors' Records, June 5, 1905, p. 87, vol. 3, Conrey Papers.

117. Jennings, "Report" (1906), p. 31.

118. Edmunds and Sweeny to Jennings, July 15, 1917, Jennings MSS.

119. Edmunds to Jennings, August 11, 1917, Jennings MSS.

120. Edmunds to Jennings, August 11, 1917, Jennings MSS.

121. Jennings to Edmunds, August 16, 1917 (copy), Jennings MSS.

122. Edmunds to F. F. Stanley, November 30, 1917, Jennings MSS; *Madisonian Times,* June 14, 1918.

123. Edmunds to Jennings, December 13, 1918, vol. 29, Conrey Papers; *Madisonian Times,* December 20, December 27, 1918.

124. Edmunds to Stanley, December 30, 1918, vol. 29, Conrey Papers; Edmunds to Charles Janin, December 26, 1918, Box 3, Charles Janin MSS. See also Pierce C. Mullen and Michael L. Nelson, "Montanans and 'The Most Peculiar Disease': The Influenza Epidemic and Public Health, 1918-1919," *Montana, the Magazine of Western History* 37 (Spring 1987): 50-61.

125. Edmunds to Stanley, September 27, 1918, telegram, vol. 29, Conrey Papers.

126. Edmunds to Stanley, October 14, 1918, vol. 29, Conrey Papers.

127. *Madisonian Times,* November 15, 1918, February 21, 1919.

128. *Alder Gulch Times,* February 17, August 11, December 22, 1905, August 20, 1915.

129. *Alder Gulch Times,* April 28, 1905, July 30, August 13, 1915.

130. *Alder Gulch Times,* May 15, 1908, May 12, 1911.

131. *Alder Gulch Times,* June 29, July 6, 1906.

132. *Alder Gulch Times,* March 6, 1908.

133. *Alder Gulch Times,* November 15, 1907, October 30, 1908.

134. *Alder Gulch Times,* October 27, 1907, April 17, 1908.

135. *Alder Gulch Times,* September 17, 1915.

136. *Madisonian Times,* November 3, November 10, 1916.

137. *Madisonian Times,* July 16, 1915.

Chapter Eight: A Mine Is a Wasting Asset

1. Jennings to H. G. Perkins, July 24, 1913 (copy), Jennings MSS.

2. Certificate of Reduction of Capital Stock, Conrey Place Mining Company, February 15, 1913, filed February 20, 1913, Secretary of State's Office, Helena.

3. Certificate of Capital Stock, Conrey Placer Mining Company, July 19, 1915, filed July 27, 1915, Secretary of State's Office, Helena; deed of July 23, 23, 1915, from Poor Farm Placer Mining Company to Conrey Placer Mining Company, Deeds Record Book LXXXIII, pp. 145-147, Madison County Courthouse.

4. The company had also paid $118,800 in dividends on Poor Farm shares, $321,812.69 on dividends on Conrey shares, $42,868.42 on interest on Poor Farm notes, and $56,813.90 on interest on Conrey notes. See Stanley to Jennings, October 21, 1915 (copy), letterpress book, vol. 16, Conrey Papers.

5. Amended Certificate of the Proceedings of the Conrey Placer Mining Company with Reference to the Reduction of the Capital Stock of Such Company, October 12, 1915, filed March 17, 1917, Secretary of State's Office, Helena.

6. Meeting of May 14, 1915, Directors' Records, p. 35, vol. 4, Conrey Papers; *Who Was Who in America,* vol. 3 (Chicago: Marquis Company, 1960), pp. 13-14; *The National Cyclopedia of American Biography,* vol. 53 (New York: James T. White & Company, 1971), pp. 11-12.

7. Thomas H. Leggett to Jennings, May 16, 1917, vol. 44, Conrey Papers.

8. Jennings to Charles F. Adams, May 10, 1917, vol. 44, Conrey Papers.

9. Edmunds, Sweeny, and Prier to Jennings, August 8, 1917, Jennings to Edmunds, August 16, 1917 (copy), Jennings MSS.

10. Meeting of September 21, 1917, Directors' Records, p. 123, vol. 4, Conrey Papers; Jennings to Edmunds, January 21, 1918 (copy), Jennings MSS; *Report of a Joint Committee Appointed from the Bureau of Mines and the United States Geological Survey by the Secretary of the Interior to Study the Gold Situation,* U.S. Bureau of Mines Bulletin No. 144 (Washington, D.C., 1919), pp. 72-73. See also Hennen Jennings, *The Gold Industry and Gold Standard* (n.p., 1918).

11. Jennings to Edmunds, January 29, 1918 (copy), Jennings MSS.

12. Jennings to Edmunds, January 21, 1918 (copy), Jennings MSS.

13. Edmunds to Stanley, April 5, 1918 (copy), Jennings MSS.

14. Edmunds to Jennings, March 13, 1918, vol. 29, Conrey Papers.

15. Edmunds to Stanley, May 4, 1918, vol. 29, Conrey Papers.

16. Edmunds to Jennings, May 9, 1918 (copy), vol. 29, Conrey Papers; *Madisonian Times,* May 31, 1918.

17. Edmunds and Sweeny to Stanley, September 15, 1918, vol. 29, Conrey Papers.

18. *Directory of Montana Metal and Coal Mines* (Butte: State School of Mines, 1919), p. 13.

19. Jennings to Stanley, March 4, 1919, vol. 30, Conrey Papers.

20. Edmunds to Stanley, July 16, 1919, vol. 30, Conrey Papers.

21. George J. Carr to Jennings, May 26, 1919, vol. 30, Record of Meeting of June 23, 1919, Directors' Records, pp. 174-175, vol. 4, Conrey Papers.

22. Edmunds to Stanley, December 3, 1919, vol. 30, Conrey Papers.

23. Edmunds to Jennings, January 19, 1920, vol. 31, Conrey Papers.

24. Meeting of December 23, 1919, Directors' Records, pp. 178-179, vol. 4, Conrey Papers.

25. Meeting of September 3, 1920, Directors' Records, pp. 183, 185, vol. 4, Conrey Papers.

26. Edmunds to Stanley, September 14, 1920, vol. 31, Conrey Papers.

27. Meeting of December 31, 1920, Directors' Records, pp. 194-195, vol. 4, Conrey Papers.

28. Meeting of November 9, 1921, Directors' Records, vol. 5, Conrey Papers.

29. Meetings of June 7, September 13, November 9, 1921, Directors' Records, vol. 5, Conrey Papers.

30. Meetings of July 10, November 14, 1922, Directors' Records, vol. 5, Conrey Papers; Statement for Dissolution, November 13, 1922, certified January 4, 1923, Secretary of State's Office, Helena.

31. Statement, Retirement of Capital Stock, Final Dividend in Liquidation, March 21, 1917, vol. 40, Conrey Papers.

32. Edmunds and Sweeny to Stanley, September 15, 1918, vol. 29, Conrey Papers.

33. Jennings to Stanley, October 14, 1918, vol. 29, Conrey Papers.

34. Ibid.; Stanley to Edmunds, October 4, 1918 (copy), vol. 7, Conrey Papers.

35. Meeting of December 9, 1918, Directors' Records, pp. 161-162, vol. 4, Conrey Papers.

36. Albert Hanford to Edmunds, January 28, 1918 (copy), Edmunds to Hanford, January 31, 1918 (copy), Jennings MSS.

37. Jennings to Edmunds and Sweeny, November 13, 1918 (copy), Edmunds and Sweeny to Jennings, November 2, 1918, vol. 29, Conrey Papers.

38. Edmunds to Stanley, December 31, 1918, vol. 29, Conrey Papers.

39. Charles Janin to Chester Purington, March 5, 1919 (copy), Charles Janin MSS, Huntington Library, San Marino.

40. Charles Janin to Edmunds, April 21, 1919 (copy), vol. 30, Conrey Papers.

41. Meeting of June 23, 1919, Directors' Records, pp. 171-172, vol. 4, Conrey Papers.

42. George J. Carr to Jennings, May 26, 1919, vol. 30, Conrey Papers; Edmunds to Charles Janin, August 2, 1919, Box 3, Janin MSS.

43. Edmunds to Jennings, November 19, 1919, vol. 30, Conrey Papers; Edmunds to Stanley, April 10, December 9, 1920, vol. 31, Conrey Papers; Edmunds to Charles Janin, November 13, 1919, Box 3, Janin MSS.

44. Edmunds to Stanley, December 9, 1920, vol. 31, Conrey Papers.

45. Meeting of November 9, 1921, Directors' Records, vol. 32, Conrey Papers.

46. Edmunds to Stanley, December 28, 1920, vol. 31, Conrey Papers. Dismantling of No. 1 at Adobeville was completed in mid-January 1921. See *The Madisonian,* January 14, 1921.

47. Meeting of September 13, 1921, Directors' Records, vol. 5, Conrey Papers.

48. Edmunds to Stanley, June 22, 1921, vol. 32, Conrey Papers; Edmunds to Charles Janin, June 17, 1921, Box 3, Janin Papers; Janin to Edmunds, June 11, 1921 (copy), Box 8, Janin MSS.

49. Meeting of November 9, 1921, Directors' Records, vol. 5, Conrey Papers.

50. Edmunds to Arthur H. Brooks, November 4, 1921, vol. 32, Conrey Papers.

51. Edmunds to Charles Francis Adams, August 2, 1921, vol. 32, Conrey Papers.

52. Machine Record Book No. 2, p. 203, Marion Steam Shovel Division.

53. Machine Record Book No. 2, p. 338, Marion Steam Shovel Division; H. Fies to Charles Janin, December 15, 1922, Box 17, Janin MSS; Cheney, *Names on the Face of Montana,* p. 233. According to the U.S. Geological Survey, the dredge worked two months in 1922 and worked 545,400 cubic yards of gravel before the equipment was sold. See *Mineral Resources of the United States, 1922,* p. 478. See also *Madisonian Times,* May 31, 1918, April 18, 1919; *The Madisonian,* April 21, May 26, June 16, 1922.

54. Edmunds to Jennings, May 5, 1918, Jennings MSS.

55. Edmunds to Stanley, June 27, 1921, vol. 32, Conrey Papers; Edmunds and Sweeny to Jennings, February 28, 1918, Jennings MSS; deed of March 24, 1919, from Conrey Placer Mining Company to Peter H. Anderson, Deed Record Book C, pp. 470-471, Madison County Courthouse.

56. Under tax sale, Frank Carey bought two acres of Conrey land for 85 cents. Deed of July 20, 1939, from Conrey Placer Mining Company to Frank Carey (Tax Sale Certificate No. 1178), Deed Record Book CXXV, p. 147, Madison County Courthouse.

57. Deed of February 25, 1921, from Conrey Placer Mining Company to Samuel B. Page, Deed Record Book CIII, p. 308, deed of November 14, 1922, from Conrey Placer Mining Company to Madison County, Deed Record Book

CVI, pp. 203-204, deed of November 20, 1922, from Conrey Placer Mining Company to Charles F. Adams, Deed Record Book CXXVI, p. 213, Madison County Courthouse.

58. For representative samples, see deed of December 27, 1920, from Conrey Placer Mining Company to Albert and Nellie P. Nelson, deed of May 31, 1921, from Conrey Placer Mining Company to Fred and Agnes Taylor, deed of December 20, 1920, from Conrey Placer Mining Company to James A. Burton, deed of July 8, 1921, from Conrey Placer Mining Company to T. E. Castle, Deed Record Book CIII, pp. 342-343, 350-351, 360-361, 372; deed of May 3, 1922, from Conrey Placer Mining Company to H. A. and Alice Petit; deed of August 16, 1922, from Conrey Placer Mining Company to Lillian Griffin, Deed Record Book CVI, pp. 56-57, 138; deed of May 19, 1922, from Conrey Placer Mining Company to A. G. Basden, Deed Record Book CX, p. 119, Madison County Courthouse.

59. Deed of November 20, 1922, from Conrey Placer Mining Company to Charles F. Adams, deed of November 13, 1922, from Conrey Placer Mining Company to Leslie C. Gilman, Deed Record Book CVI, pp. 213, 232-233, Madison County Courthouse. As of the fall of 1989, Leslie's son, Lowell Gilman, and his family still lived on the property.

60. Jennings, *The McKay Endowment and Applied Science at Harvard,* p. 10. An undated share statement after McKay's death shows that the Conrey Placer Mining Company had a total of 8,606 shares—3,000 common, 3,000 First Preferred, and 2,606 Second Preferred. Of these, the McKay estate owned 1,170 common, 2,590 First Preferred, and 1,070 Second Preferred—a total of 4,830, more than a majority interest of 4,304. Share statement, n.d., vol. 40, Conrey Papers.

61. Statement, Retirement of Capital Stock, Final Dividend in Liquidation, March 21, 1927, vol. 40, Conrey Papers.

62. Conrey Placer Mining Company Yield and Yardage Record, 1897 to 1920, vol. 41, Conrey Papers.

63. *The Madisonian,* March 4, 1955. Contemporary sources estimated that the Conrey Company had spent over $1 million on dredges and other equipment and that it took out $10 million in gold. See *The Madisonian,* January 27, 1922.

64. Conrey Placer Mining Company Yield and Yardage Record, 1897 to 1920, vol. 41, Conrey Papers; Conrey Placer Mining Company Dividends, vol. 40, Conrey Papers; *Great Falls Tribune,* June 13, 1971. Efforts to locate this film in various federal depositories have been unsuccessful.

65. Conrey Placer Mining Company Dividends, vol. 40, Conrey Papers.

66. *The Mines Handbook,* vol. 15 (Tuckahoe, New York: Mines Handbook Company, 1922), p. 1058.

67. *Montana Standard,* March 14, 1971; Aubury, *Gold Dredging in California,* p. 254; Thomas T. Read, *The Development of Mineral Industries Education* (New York: American Institute of Mining Engineers, 1941), p. 143; *Harvard Alumni Bulletin* 52 (November 5, 1949); 155-156.

68. Jennings, *History of Dredging,* p. 9. I am indebted to Howard A. "Bud" Packard of Great Falls for his compilation of a listing of Montana's dredges for this period.

69. *Thirteenth Report of the Bureau of Agriculture, Labor and Industry of the State of Montana for the Years 1911 and 1912* (Helena: Independent Publishing Company, 1913), p. 169.

70. Albert E. Humphreys, Jr., to D. Holmes Morton, December 12, 1934 (copy); *The Rocky Mountain News,* December 1932, clipping, "Notice of Special Meeting of the Stockholders of Humphreys Gold Corporation," Denver, April 23, 1935, Box 12, Box 45, Albert E. Humphreys MSS, Colorado Historical Society, Denver; Charles Janin to Walter Johnson (Twin Bridges, Montana, May 27, 1936 (copy), Box 9, Janin MSS; Leyden, *Gold Placers of Montana,* p. 82; Arthur V. Corry, "Dragline Installation for Recovering Gold at Virginia City, Mont.," *Mining and Metallurgy* 17 (October 1936): 468-470.

71. Dick Pace, *Golden Gulch: The Story of Montana's Fabulous Alder Gulch* (n.p., 1962), p. 77; Federal Writers' Project, *Montana: A State Guide Book* (New York: Hastings House, 1939), p. 359.

72. Michael J. Penfold, *Historical Comparison Photography: Mountain Foothills Dillon Resource Area Montana* (n.p.: Bureau of Land Management, n.d. [after 1979]), pp. 62-63.

73. Olin D. Wheeler, *Wonderland 1902* (St. Paul: W. C. Riley, 1902), p. 35.

74. *The Madisonian,* January 27, 1922.

75. *Alder Gulch Times,* December 11, 1908, November 17, 1911.

76. See Spence, "The Golden Age of Dredging: The Development of an Industry and Its Environmental Impact," *Western Historical Quarterly* 11 (October 1980); 409-414; Don Chaput, "Dredging's Golden Age: Did the Fox Guard the Chickens?" and Clark C. Spence, "Brer Fox in the Hen House: A Reply" *Western Historical Quarterly* 12 (July 1981): 305-307.

Bibliographical Essay

The real core of this study comes from several major manuscript collections that deal with the Conrey operations in considerable detail. Of primary importance are the forty-four volumes of letters, documents, records, and photographs in the Baker Library, Harvard School of Business, pertaining to the German Bar Company, the Conrey Placer Mining Company, and the Poor Farm Mining Company and nearly all aspects of their activities, financial and operational. Also of more than passing significance are the Hennen Jennings manuscripts, located in the American Heritage Center of the University of Wyoming, which houses the greatest collection of mining manuscripts in the country. Jennings was a typical mining engineer, with a career that took him to the ends of the earth. But for a dozen years or so he served as vice-president and consulting engineer for the Conrey Company; his thoughtful reports after on-the-spot visits to Alder Gulch and his conscientious correspondence with the managers at Ruby help us to understand and to put into perspective the typical problems and solutions of an early dredging enterprise. Files of the Marion Power Shovel Division of Dresser Industries, successor of the Marion Steam Shovel Company, of Marion, Ohio, give a wealth of technical information on specifications and the construction of electric dredges No. 1, No. 2, and No. 4.

More official sources have also been useful, especially the Madison County records located in the courthouse in Virginia City and the files of the Montana Secretary of State's Office in Helena. Lowell Gilman of Ruby, son of veteran clean-up and retort man, Leslie Gilman, owns a truly remarkable photograph collection, with hundreds of pictures depicting almost every aspect of the Conrey operations, from the Lidgerwood cableway, through all the steam and electric dredges.

Newspaper information is scattered, but most important has been the run of the *Alder Gulch Times* of Virginia City, with name changes to *The Madisonian* and *Madisonian Times.* Ruby and Conrey were on the fringe, but they were never ignored by the county seat newsmen. Also worthwhile have been files of *The Dillon Examiner.*

No study of mining in any part of the United States could be completed without recourse to the standard periodicals of the industry, *The Engineering and Mining World* of New York and *Mining and Scientific Press* of San Francisco, which merged in 1922. The frequent, detailed articles and letters in both indicate the effect the Conrey Placer Mining Company had among mining men of the three decades following 1897. Other industry organs, including *Engineering News, Mining Science, Mining and Metallurgy,* and *Mining and Engineering World,* have also been useful.

Among Montana state publications, the reports of the Bureau of Agriculture, Labor and Industry from 1902 to 1912 have been of use, as has the general survey, Charles J. Lyden, *The Gold Placers of Montana,* Montana Bureau of Mines and Geology Memoir No. 26 (Butte: Montana School of Mines, 1948). Likewise, Lewis E. Aubury, *Gold Dredging in California,* California State Mining Bureau Bulletin No. 57 (Sacramento: W. W. Shannon, 1910), gives both specific and background information.

Several United States Bureau of Mines bulletins have been invaluable. The most significant is Hennen Jennings, *The History and Development of Gold Dredging in Montana,* U.S. Bureau of Mines Bulletin No. 121 (Washington, D.C., 1916), which is essentially the Conrey story. Broader, but also of utmost concern to any historian of dredging in Alder Gulch is Charles Janin, *Gold Dredging in the United States,* U.S. Bureau of Mines Bulletin No. 127 (Washington, D.C., 1918), which draws on Jennings but sets the Conrey enterprise against a broader backdrop.

Always good for biographical insight about the mining world is Thomas A. Rickard, *Interviews with Mining Engineers* (San Francisco: Mining and Scientific Press, 1922). *The Autobiography of Nathaniel Shaler* (Boston: Houghton Mifflin, 1909) and Walter Louis Berg, "Nathaniel Southgate Shaler: A Critical Study of an Earth Scientist" (Ph.D. diss., University of Washington, Seattle, 1957), add another dimension about two of the organizers of the company, Shaler and Gordon McKay. Unfortunately, there are no general histories of American mining that do more than mention dredging in a passing fashion, let alone details of operations in Alder Gulch.

Index

Accidents, 73, 75, 81, 96, 97: settlement after, 96, 97; company liability for, 98
Adams, Charles Francis II, 49, 89, 100, 106, 112-113, 116, 120
Admiralty Zinc Company, 19
A. H. Cox & Company, 118, 119
Agassiz, Louis, 21
Alaska, 10, 11, 12, 71, 123, 124
Alcohol consumption, among workers, 90, 91, 111
Alder, 43, 94
Alder Gulch, 13, 15, 17, 20, 21, 25, 27, 28, 30, 33, 45, 49, 55, 57, 59, 73, 81, 82, 83, 84, 88, 107, 114, 118, 124-125
Alder Gulch Consolidated Mining Company, 83
Alder Gulch Times, 34, 35, 63, 71, 96, 100, 108
American Bridge Company, 42
American Manganese Steel Company, 77
American Smelting and Refining Company, 78
Anderson, Thomas, 98
Baier, Julius, 42, 46, 54, 55, 82, 88
Baker, Charles S., 111
Bannack, Montana, 8, 9, 10, 35, 36, 39, 52, 123
Barker, John G., 118
Barton Gulch, 84
Batten, John, 84
Belt stacker, 10, 61
Bigelow, Henry M., 20, 21, 22, 23, 24, 53, 121
Bigelow, Jeanette, 26
Bigelow, Walter, 118, 121
Bismarck-Nugget Gulch Company, 46
Bixby, Harry G., 20, 21, 22, 23, 24, 39, 53
Bixby Mining Association, 22, 23, 26
Bon Accord, 8
Brenner, Howard H., 60, 64
Brissette, Nelson A., 94
British Institute of Mining and Metallurgy, 48
Brook, Lawrence, 107
Brooks, T. B., 19
Brynie, Axtel, 102, 103, 119
Buck, M. E., 64
Bucket chain, 13, 48, 55
Bucket line, 9, 14: improvements in, 14
Bucket-and-ladder dredge, 6, 29
Buckets, 70: types of, 7, 30, 31, 50, 55
Bucyrus Company, 8, 11, 30, 36, 39
Buford, Joe, 119
Bureau of Land Management, 124
Burroughs, Ambrose H., Jr., 91, 115
California-type dredges, 5, 9, 15, 50, 60, 122
Carr, George J., 15, 118
Chicago Bridge & Iron Company, 42
Chicago Drainage Canal, 30, 35
Chicago Mining and Development Company, 35, 36
Chinese miners, 26
Clancy, Dr. D. F., 109
Clark, William Alexander, 22, 84, 90, 98, 108, 121
Clean-up crews, 93
Coast Company, 8
Collins Ranch, 23, 26, 53, 108, 120
Colorado School of Mines, 89

Conrey, Philip, 24
Conrey, Sarah V., 24
Conrey Placer Mining Company, 2, 12, 15, 17, 22, 26, 45, 47, 54, 88, 92, 98, 106, 116, 120, 123: acquisition of property by, 24; as part of revolution, 12; assets of, 113, 117-119; debts of, 113; delays in operations, 77; depleting deposits of, 114; directors for, 87; dissolution of, 117; economic impact of, 80; effects of World War I on, 100, 110; employee relations of, 99; formative years of, 19; housing for, 107-108; employees for, 79; innovation by, 14; labor for, 95, 106; layoffs by, 115; legal disputes, 97, 116; liquidation of, 120; litigation, lack of, 121; longevity of, 83; lost time record for, 73, 80; operating costs of, 48, 84, 103, 115; profits of, 32, 122; relations with town, 110; road work for, 79; safety precautions taken by, 99; sale of real estate, 120; stock in, 23, 113, 116, 117, 121, 122; taxes paid by, 116, 124; testing by, 82; total extractions of, 121
Conrey Ranch, 20, 107
Contracts, 40, 62
Cordell, Charley, 97
Couch, Thomas, 7
Cox, Madge, 64
Crosser, J. G., 111
Cyanidation, 3, 48
Deadmen, 40
Depreciation, 39, 88
Dessault, Leon, 96, 98
Dionne, Joe, 98
Doherty, H. L., 96
Dredges, 3, 5, 6, 48, 92, 121: cost of, 4; crew needed for, 92; effects of on the land, 4; failure of, 82; in Alaska, 10, 11, 12, 71, 123, 124; in California, 5, 7, 10, 18, 19, 48, 50, 51, 123, 124; in Colorado, 10, 19, 123; in Idaho, 10; in Montana, 10; names for, 4, 111; number of, 6 technology of, 13; types of, 4-11 29, 35, 36, 43, 50, 60
Dredges, electric, 7, 11, 13, 35, 39, 40 57, 59, 80, 99, 107, 121: engines for 10, 31; statistics for, 44
Drills, 82
Edmunds, George, 70, 78, 79, 84, 88, 90, 91, 93, 94, 100, 101, 107, 108, 110, 111, 115, 116, 119
Electric hoists, 64
Electrical power, 46, 48, 50
Eliot, Charles W., 49
Elling, Henry, 25
Elling, Mary, 25
Elling Ranch, 25
Elphison, Frankie, 111
Employees, 101, 104-106: death of, 96, 97, 98, 109, 110, 111, 116, 118, 122; descriptions of, 93; entertainment for, 89, 91, 111; funerals for, 92; marital status of, 107; wages for, 92, 99, 100, 101, 103, 104, 105, 106; women, 103
Experimentation, 12, 14, 17
Fielding L. Graves, 7, 123
Fire, 69, 70, 78, 79
Flooding, 72
Foster, C. C., 96
Fuel, 10, 78
Fulton Iron Works, 41
Gavin, C. D., 23, 27
General Electric, 66, 82
German Bar, 20, 23, 26, 29, 30, 32, 35, 40, 81
German Bar Fluming Company, 25, 29
German Bar Mining Company, 23, 26, 30
Gerry, M. H., 89
Gilman, LeRoy, 112
Gilman, Leslie, 83, 94, 98, 119, 120, 122
Gold: as a secondary metal, 12; price of, 12; production levels of, 10
Gold Dredging Company, 8
Gold mining, during World War II, 12

Gold-saving tables, 9, 116
Goodspeed, William, 111
Gordon (No. 1 steam) 1, 40, 41, 44, 45, 46, 50, 51, 57, 60, 71, 72, 73, 74, 75, 88, 92, 97: cost of, 41; rebuilding of, 41; the sinking of, 76
Groman's Columbia Gardens Orchestra, 110
Graeter, Augustus F., 8
Grasshopper Creek, 7, 8, 35, 36, 123
Graves, Fielding L., 8, 35
Gunther, John, 4
Hagenach, Jacob J., 98
Hammon, Wendell P., 7, 49, 59, 60
Hanford, Albert, 117
Harper, Samuel S., 8
Harvard University, 1, 17, 22, 47, 112-113, 122
Hayward "orange peel" shovels, 29, 33, 34, 121
Hayward Company, 33, 34: buckets made by, 33, 41
Hildreth, W. O., 42
Hoffman, Ross, 84
Holmes, George L., 50, 59
Hulls, 10
Humphrey, Albert E., 123
Humphrey, Ira Boyd, 123
Humphreys Gold Production Corporation, 123-124
Hutchins, John Power, 13, 14, 52, 72
Hydraulic mining, 8
Influenza epidemic, 109
Injuries, 31
Innovation, 12, 121
Insurance, 79, 81, 88, 97, 98, 99, 115, 116
Isolation, 15
International Workers of the World (IWW), 95, 106
Jackling, Daniel C., 3
Janin, Charles, 11, 116, 119
Jennings, Coleman, 121
Jennings, Henning, 1, 13, 14, 42, 47, 55, 58, 59, 62, 64, 68, 70, 75, 79, 80, 87, 89, 90, 92, 93, 99, 100, 101, 103, 104, 108, 113-114, 116, 121
Johanson, Edward, 63
Jones, Dr. Arthur O., 109
Kammerer, Charles, 51, 58, 70, 83, 89, 92, 95, 97, 98, 101, 104, 110, 120
Keystone drill, 82, 83, 84
Kincaid, McQueen & Company, 6
King, David H., 119
King, George W., 64
Knight excavating bucket, 30, 31
Kronwall, Olaf, 103, 115-119
La Grange Gold Dredging Company, 118, 119
Lakes, Arthur, Jr., 3
Leggett, Thomas H., 90, 114
Legree, Arthur, 95
Lena Creek Mining and Dredging Company, 119
Lidgerwood cableway system, 29, 31, 32, 33, 40
Lidgerwood Manufacturing Company, 30
Lidgerwood Radial Cableway, 30
Linemen, 93
Louisiana Purchase Exposition, 45
Machine shops, 50, 75, 78, 98, 102, 120: investment in, 70
Madison County Poor Farm, 23, 26, 53, 107
Madison River Power Company, 47, 64, 66, 81, 124
Maggie A. Gibson, 8, 14, 35, 37, 39, 40, 41, 42, 45, 46, 48, 79, 92: installed on Alder Gulch, 35; launched at Bannack, 36
Malone, Fred, 95
Manganese steel, 14, 67: discovery of, 9
Marion Steam Shovel Company, 11, 30, 54, 59, 60, 62, 63, 97, 101, 120
Mass production, 3, 4, 18
McKay, Gordon, 1, 21, 22, 23, 24, 30, 39, 40, 44, 47, 54, 60, 87, 122
Medical care, 108
Mein, W. W., 70
Mercury amalgam, 62, 72, 93
Metlin, George, 8
Midvale Steel Company, 41

Mine tailings, 6, 57
Moenster, John, 97
Monaghan, J. H., 97
Montana Bureau of Agriculture, Labor and Industry, 45
Montana Flume and Mining Company, 23, 25
Montana Power Company, 12, 47, 81, 102, 115
Montana Workman's Compensation Act, 99
Moore, Eva, 26
Moore, Philip N., 19, 20, 21, 22, 23, 26, 29, 30, 32, 39, 42, 47, 53, 87, 88
Mortgage bonds, 24
Mountaineer Welding Company, 96
Multiple-dredge concerns, 12
Myers, James J., 47, 48, 53
National Union Insurance Company, 58
Natomas Company, 14, 109
Nelson, Albert, 97
New York Engineering Company, 11
New Zealand, dredging in, 3, 5, 6, 7, 8, 9, 35
New Zealand Engineering Company of Dunedin, 7
No. 1 electric, 59, 61, 62, 72-73, 74, 76, 78, 79, 80, 81, 90: repair of, 93, 94, 97, 103, 114, 115, 116, 118-119: dismantling of, 122
No. 2 electric, 59, 60, 61, 62, 70, 71, 74, 76, 80, 83, 93, 94, 97: remodeling of, 104, 115, 116, 118, 119: dismantling of, 122
No. 3 electric, 52, 54, 55-58: description of, 59: drydocked, 72, 73, 74, 75, 76, 78, 79, 81, 89, 93, 96, 98, 100, 114, 115, 118, 119: dismantling of, 120
No. 4 electric, 14, 54, 62-64: construction of, 63: shipping of parts and material and cost of, 64: start up date, 64-67: description of, 66-68: capacity inadequacy and repair, 70, 73, 74, 79, 81, 83, 88, 91, 93, 96, 98, 104-105: strike, 106, 115, 116, 119, 120: scrapping of, 122, 123: retirement of
No. 1 steam. See *Gordon*
No. 2 steam, 42, 43: delivery and reconstruction of, 44, 45, 46, 50, 51, 57, 60, 72, 73, 75, 88, 89, 96
Northern Pacific Railroad, 26, 43, 78, 110, 124
Northwest Steel & Metal Products, 120
Oilers, 93
Old Gordon. See *Gordon*
Oregon Shortline Railroad, 124
Payne, W. W., 83
Pearson, Olaf, 103
Perry, O. B., 60
Pilger, Newton, 83
Pilotmen, 93
Pioneer Gold Dredging Company, 119
Postlethwaite, Robert H., 7, 9, 35
Poor Farm Placer Mining Company, 2, 26, 53, 62, 63, 113: stock in, 53
Prier, Truman D., 64, 81, 89, 90, 108, 110
Priestman "grab," 34
Processing, description of, 61
Prospecting, 91
Pumpelly, Raphael, 19
Reclamation, 120, 124
Reconstruction Finance Company, 123
Reiling, Herman J., 8
Repairs, 74, 79, 94
Retorting, 3: house for, 50, 120; furnaces, 94; shops, 50, 69, 70, 75, 78, 94, 98, 102, 120
Revett, Benjamin, 8
Risdon Iron Works, 7, 11, 35, 39, 40
Royalties, 36
Ruby, 18, 46, 52, 55, 81, 87, 88, 92, 99, 100, 106-107, 110, 111, 124: businesses in, 107, housing in, 107; medical care in, 108
Ruby Hall, 110
Ruby Hotel, 108
Ruby Post Office, 120

Ruby River, 17
Schools, 107
Scott, Adeline, 111
Sculling-Gallagher Iron & Sheet Company, 41
Sedman, Oscar A., 25
Seward Peninsula, 10
Shaler, Nathaniel S., 1, 19, 20, 22, 23, 24, 26, 30, 46, 47, 48, 53, 84, 87, 89, 121
Shaler, Page, 121
Shott, Charlie, 102, 108
Slate Creek Iron Property, 19
Sloss, John C., 20, 23, 29, 30, 82
Sluices, 6, 29, 31, 44, 61, 71
Smith, E. M., 95
Sparrow, R. S. & Company, 6
Spuds, 36: aerial, 14; description of, 61; steel-pointed, 9
Stanley, Frank F., 42, 48, 53, 79, 88, 99, 100, 106, 113, 115, 116, 121
Stanley Manufacturing Company, 42, 43
Steam heat, 58, 62
Steampowered shovels, 30
Strikes, 99, 100, 104, 105
Strouse, Fred, 102
Surveying, 83, 91, 114
Swank Creek Mining and Dredging Company, 119
Sweeny, Oscar O., 90, 91, 92, 93, 97, 101, 110, 117, 120
Tanner, A. W., 84
Technicians, 7
Test drilling, 114
Timber, costs of, 10
Townsend, Dr., 84, 108
Tranffler, Nicholas, 95
Trommels, 9, 13
U.S. Army Corps of Engineers, 92
U.S. Assay Office, 78
U.S. Smelting, Refining & Mining Company, 10, 12
Underground workings, 78
Union Construction Company, 11, 101
Virginia City, Montana, 12, 18, 22, 108, 124
Vulcan Iron Works Company, 39, 40
Ward, W. F., 89
Western Engineering & Construction, 11
Westinghouse Electric and Manufacturing Company, 55, 67, 81, 82
Wheeler, Olin D., 124
Wiles, Herbert, 98
Winching machinery, 65
Winter operations, 10, 14, 58, 69, 71, 83, 116
Wiseman, Adolph H., 14, 49, 51, 89, 99, 110
Worthington centrifugal pumps, 66
Yuba Construction Company, 59
Yuba Manufacturing Company, 11